Study Guide
for
Sociology
SIXTH EDITION

by Rodney Stark

Carol A. Mosher
Jefferson Community College

Wadsworth Publishing Company

I(T)P™   An International Thomson Publishing Company

Belmont • Albany • Bonn • Boston • Cincinnati • Detroit • London
Madrid • Melbourne • Mexico City • New York • Paris • San Francisco
Singapore • Tokyo • Toronto • Washington

Printed in the United States of America
1 2 3 4 5 6 7 8 9 10—01 00 99 98 97 96

For more information, contact Wadsworth Publishing Company.

Wadsworth Publishing Company
10 Davis Drive
Belmont, California 94002, USA

International Thomson Editores
Campos Eliseos 385, Piso 7
Col. Polanco
11560 México D.F. México

International Thomson Publishing Europe
Berkshire House 168-173
High Holborn
London, WC1V 7AA, England

International Thomson Publishing
GmbH
Königswinterer Strasse 418
53227 Bonn, Germany

Thomas Nelson Australia
102 Dodds Street
South Melbourne 3205
Victoria, Australia

International Thomson Publishing Asia
221 Henderson Road
#05-10 Henderson Building
Singapore 0315

Nelson Canada
1120 Birchmount Road
Scarborough, Ontario
Canada M1K 5G4

International Thomson Publishing
Japan
Hirakawacho Kyowa Building, 3F
2-2-1 Hirakawacho
Chiyoda-ku, Tokyo 102, Japan

ISBN 0-534-25714-3

CONTENTS

Preface   Studying "Smarter"—How to Make the Most of This Study Guide

Introduction   Studying a Text: The SQ4R Method 1
Effective Test Taking 4
Essay Questions: Understanding Format 7

Chapter 1   Groups and Relationships: A Sociological Sampler 9
Chapter 2   Concepts for Social and Cultural Theories 19
Chapter 3   Micro Sociology: Testing Interaction Theories 29
Chapter 4   Macro Sociology: Testing Structural Theories 38
Chapters 1–4   Review and Special Project 47
Chapter 5   Biology, Culture, and Society 48
Chapter 6   Socialization and Social Roles 57
Chapter 7   Crime and Deviance 67
Chapter 8   Social Control 77
Chapters 5–8   Review and Special Project 86
Chapter 9   Concepts and Theories of Stratification 87
Chapter 10   Comparing Systems of Stratification 96
Chapter 11   Intergroup Conflict: Racial and Ethnic Inequality 105
Chapter 12   Gender and Inequality 114
Chapters 9–12   Review and Special Project 123
Chapter 13   The Family 124
Chapter 14   Religion 133
Chapter 15   Politics and the State 142
Chapter 16   The Interplay Between Education and Occupation 151
Chapters 13–16 Review and Special Project 161
Chapter 17   Social Change and Modernization 162
Chapter 18   Population Changes 171
Chapter 19   Urbanization 180
Chapter 20   The Organizational Age 189
Chapter 21   Social Change and Social Movements 198
Chapters 17–21 Review and Special Project 207

Studying "Smarter"

How to Make the Most of This Study Guide

This study guide is designed to aid your study of the sixth edition of Rodney Stark's *Sociology*. It is meant to complement the text, not substitute for it. The text is well written and contains a common thread that ties the chapters together; you will find it informative and lively reading. But even with a well-written text, students sometimes find it difficult to effectively focus their study and ferret out the most important information. This study guide will assist you in this process.

There is no foolproof way to study and no one way to use this study guide. This guide might be used in any (or all) of the following ways:

1. Before you read each chapter, the study guide can alert you to the key concepts, theories, and research studies discussed in the chapter. This preview will enable you to focus your attention effectively.

2. Throughout your reading and study, it can reinforce your knowledge of important topics.

3. After you complete each chapter, it can serve as a review and self-test to ensure that you understand the most important material.

4. Prior to an exam, it can serve as a review and refresher, alerting you to areas that may need further study.

You will want to develop a system that works best for you. Your professor or other students may be able to offer additional suggestions for effective use.

Format of the Study Guide

The text is divided into five parts, each containing between two and five chapters. (Individual professors may deviate from this format when designing courses.) Chapters in the study guide are grouped together to follow a format similar to that in the text, but they allow flexibility in the event of any deviation.

The study guide's twenty-one chapters directly correspond to the chapters in the text. Each chapter contains the following:

1. An overview that briefly highlights the main topics in the chapter.

2. A capsule summary that condenses the chapter into a few paragraphs. Key topics and concepts are underlined for easy reference.

3. A list of key concepts, theories, and research studies with specific page references for key concepts and research studies. These form the basis of sociology and sociological inquiry, and a thorough knowledge of them is essential for mastery of the material.

4. Completion, or fill-in-the-blank, questions that are drawn from all aspects of the chapter. The answers appear at the end of the chapter.

5. Multiple-choice questions that cover the entire chapter. Again, the answers are at the back of the chapter.

6. Essay questions or topics that require a synthesis and application of the important material. Many of these use the key words described in the section on how to study. The first two essays in each chapter reflect the cognitive levels described in the introductory section on essay questions. A final question asks you to apply what you have learned to everyday life.

Following each group of chapters is a brief review of the chapters plus a suggestion for a special project designed to offer creative ways of actually putting sociology to use with the information gained in that section. The directions for these projects are open-ended to allow flexibility and creativity and are meant as suggestions rather than specific assignments. You should consult your instructor for advice and direction before undertaking these projects. In addition, this study guide contains a short section on study techniques, describing strategies for effective study and test taking. It is designed to be useful not only in your sociology course but also in other college courses. Many of the completion and essay questions contained in this guide apply these techniques.

Of course, study guides do have limitations. No study guide can ever substitute for a thorough reading and comprehensive study of a text. That is not the purpose of a study guide, nor should it be. Many professors cover additional material in lectures, which obviously cannot be included in this guide; similarly, outside readings are often assigned, which must be studied separately. Occasionally, an individual professor may emphasize a specific area in the text in greater depth than it is treated here. This guide should be used as a resource and study aid and not be relied on as the only tool for effective study. Nevertheless, this guide should assist you in your study and help you to appreciate the challenge and excitement of sociology.

Acknowledgments

I would like to thank Rodney Stark for asking me to write this study guide and Susan Shook and Wadsworth Publishing Company for their assistance.

The section describing the SQ4R method was written by Nancy Hoover, provost at Manatee Community College. She willingly let me include it here. Sarah Dye, of Elgin Community College, provided me with extensive material on test-taking techniques. The section on effective test taking is drawn from her material. In addition, Margo Elliott, associate professor of psychology at Columbia College, wrote the section titled "Essay Questions: Understanding Format" and helped write the essay questions. I am indebted to these colleagues for their information and assistance.

Carol Mosher

STUDYING A TEXT: THE SQ4R METHOD

Nancy Hoover
Provost
Manatee Community College

Author's Note

Success in sociology or any other college course depends in part on your ability to study effectively. Good study techniques allow you to make the most of your study time and can ultimately lead to a better understanding of course material and better grades. Although no method of study works for all students in all courses, experts on study skills have devised several techniques that have proved successful. Of course, no study technique can compensate for failure to attend classes or failure to complete assigned work. These techniques can, however, allow you to retain more information and use your study time more effectively.

Many students, particularly first-year students, simply do not know how to study. They complain that they spend many hours studying, yet they perform poorly on exams and other assignments. I can appreciate their position because, as a freshman, I spent many hours trying to memorize whole passages from texts the night before an exam only to find that I had missed much of the important information and had forgotten most of what I had memorized. True, I had spent many hours studying, but they were, for the most part, wasted. I simply did not know how to study effectively. Once I learned basic study techniques, I actually spent fewer hours studying, yet I learned more and received better grades.

This section offers strategies for effective study by describing the SQ4R method for studying texts and providing suggestions for taking objective and essay tests. If you want to improve your study skills, investigate the resources available on your campus. Many colleges and universities have study skills centers and learning resources centers that are staffed by professionals trained in study skills; other campuses use specially trained counselors, tutors, or advisors to serve this function. Often professors or teaching assistants are more than willing to help. Most of these services are free of charge to students, but you must seek them out.

Carol Mosher

When an instructor gives that famous assignment, "Read Chapter 8 for next time," what does she mean? There are two jobs for the student implicit in that assignment. The first is to understand; the second is to remember. The instructor does not care how the information is received or stored, just as long as it is.

Among several steps to understanding and remembering, the first is to be able to organize ideas into main ideas and details and then to see the relationships between and among these ideas. For instance, if you drive into a gas station and ask the mechanic to tell you what is wrong with your car, you would be unhappy if he said, "I can name all the parts of the carburetor, something I learned in my auto mechanics class." And yet this is the kind of learning many students do, mistaking it for the real thing. The problem with our mechanic is that he has concentrated on the details, thinking that they are as important as the main ideas; maybe he is unable to tell the difference between the two. This kind of learning is insufficient and does not reflect the way experts in the field think. If you are to pass tests prepared by your instructors, who are experts in their fields, you need to learn to think like them.

One excellent way to organize your textbook study is to use the SQ4R study technique with its six steps: survey, question, read, recite, rite, and review.

Survey

The purpose of the survey step is to aid understanding and increase reading speed. Research shows that students who survey before they read material read it 24 percent faster than do those who do not survey. Why? Because if you know where you are going, you get there faster.

How do you survey? Read the title. Think about it for a few seconds. What do you know about it already? What do you think the chapter will include? Next, read the introduction, which provides an overview of the chapter. Read all the headings and subheadings. Look at pictures, charts, or graphs. Read the summary; see how it mirrors the introduction. Finally, read any questions, terms, or other important material at the end of the chapter. This entire process should take not more than three to five minutes. Now return to the beginning of the chapter and start the second step.

Question

The question step helps reveal the organization of the chapter and the relationship of details to main ideas, so that you will not end up like our auto mechanic, not seeing the forest for the trees. To do the question step, simply change headings into questions. For example, if the heading is "Racism," the obvious question is "What is racism?" Or if the subheading is "The Causes of Racism," the question would be "What are the causes of racism?" To begin relating ideas to each other, you can ask questions that relate subheadings to headings. If the heading is "Racism," and the subheading is "Forms of Discrimination," a good question would be "How does discrimination lead to racism?" or "Is discrimination a necessary part of racism?" Don't these sound like good test questions? You can discover questions in headings, from study guides, from class discussion, and throughout the text.

Read

Next, read to answer your questions. This reading is much different from simply starting at the beginning of every chapter and reading every word, hoping that important ideas somehow will pop into your brain. As you look at your textbook, it is easy to answer the question, "What are the causes of racism?" The answers may be numbered, appear in boldface type, or be the first sentence in each paragraph. A glance through your text will reveal how easy it is to understand the structure of a textbook. Nearly all textbooks are put together in a similar way, so once you understand this method, you can apply it to every text.

Recite

Once you have located the answer in the text, your next job is to put that information in short-term memory so that you will be able to retrieve it later. The recite step will accomplish that for you. How do you recite? Look at the question you have written, look away, and answer the question out loud and in your own words. Answering the question out loud helps you to remember the answer; answering in your own words ensures that you understand the answer.

If you are unable to answer the question or do not understand it, mark it in your SQ4R notes so that you can listen especially for that point when it is discussed in class. You should do SQ4R with your chapter before you go to the class discussion on it. If you have not prepared for the class, you will be lost trying to cope with new ideas, new vocabulary, and disorganization all at once.

When you have answered your question, simply move on to the next heading or subheading, make up a question, and answer it out loud and in your own words. Continue this process to the end of the chapter.

Rite

Rite, a phonetic rendering of "write," means you should learn to use cue notes. A cue is a word or phrase that helps you to recall longer phrases. For example, "soc." might represent "socialization" or some similar term that is too long to write out every time it is used. "Soc?" might represent "What is the definition of socialization?"

As you form the questions and find the answers, write them down in cue form. In this way, you can have all the questions and answers from your chapter in a brief format.

Review

The review step, the final task, is to ensure that what you have understood will be retrievable from your memory one, two, or several weeks after you study it. There are two important times to review. The first is before you finish a study session. Review what you have covered in that session by repeating the recite step. Ask yourself the question, look away, and answer it. If you cannot answer the question, look back at the answer, then ask yourself the question again. Repeat this process until you can answer the question.

The second review should take place once a week until you are tested on the material; again, use the recite process. This review should take no more than five to ten minutes.

SQ4R was formulated in the early 1940s. No one has developed a better method for understanding and remembering textbook material. Research has demonstrated that the use of SQ4R will ensure an average of 80 percent retention of textbook material. Many of you will do much better than that. Try it. The proof is in the performance.

EFFECTIVE TEST TAKING

Most professors rely on periodic exams to assess students' knowledge of course material and to determine final grades. Faithful study throughout the course is a necessary prerequisite for taking exams. Nevertheless, even the best-prepared students often approach exams with some degree of anxiety and apprehension. Knowledge (and practice) of effective test-taking techniques can reduce this anxiety and increase the probability of receiving a high grade.

Exam questions typically fall into one of two categories: objective or subjective (essay) questions. Objective tests require short, specific answers. In contrast, subjective tests require broader, in-depth answers. Objective questions typically do not rely on your instructor's personal judgment to determine whether they are correct. Subjective questions, however, are often graded on both form and content and depend in part on personal judgment to determine their quality. The answer to a subjective question may be technically correct yet not be given full credit if it does not provide enough information or is organized and written poorly.

Regardless of the type of test, certain strategies for preparing for and taking exams have proved effective. Although no technique can ensure success, research has shown that when the following practices are consistently followed, they can be of great benefit to students.

Basic Strategies

Before the Exam: Preparation, of course, begins with basic study. If you have kept up with reading, class notes, and other assignments, preparation should be a matter of review rather than learning new material. The following strategies are effective as the test draws near:

1. Take full advantage of any review time available from the instructor. Clear up any questions you may have about content and find out the exact structure of the test.
2. Review your SQ4R notes. Be alert to any problem areas.
3. Make up practice questions and answer them.
4. Use a buddy system. Compare your notes and practice questions with someone else in the class. You may have missed something that he or she noticed.
5. Get a good night's sleep the night before the test.
6. Bring all necessary materials and an extra pen and pencil.
7. Arrive at the testing center a few minutes early to obtain your favorite seat and get organized.

Once the Test Has Begun:

1. Read and follow all directions. Be alert to possible choices of which questions to answer, time limits, point values, and so on.
2. Quickly read through all the questions. Budget your time.
3. Answer the questions you are sure of first. Go back to the others.

On Completion:

1. Go back and check your answers.
2. Be certain you have not inadvertently omitted any questions and that you have followed the directions.
3. Check your spelling and grammar.

Objective Tests

Objective tests typically require short, specific answers testing your ability to recognize and recall information. Types of objective questions include multiple-choice, matching, completion (fill-in-the-blanks), identification, and true or false.

When you take objective tests, it is especially important to follow specific directions because these exams are often scored by a computer or other device that has been programmed to accept only specific answers in specific places. By failing to use a pencil or making your pencil marks too light, for example, you can lose points even though your answer may be correct.

It is equally important to pay specific attention to point values of questions and sections. Objective tests often consist of a large number of questions that are worth only a few points each. Spending too much time on a question worth only two or three points can seriously jeopardize completion of the rest of the exam. In the long run, it is better to miss a few points than run out of time and be forced to omit an entire section that is worth far more.

There are techniques for taking the various types of objective tests, which you should be able to learn at your campus learning center or other resource. You will find the information you gain worthwhile.

Subjective Tests

Subjective, or essay, questions are typically more general than objective questions and rely on your ability not only to recall or recognize information but also to synthesize, organize, and explain the information in-depth. Because essay questions do tend to be broad, students often assume such questions require less study. In reality, however, they typically require more study because a deeper understanding of the material is required to answer the question successfully.

Many general test-taking strategies apply to subjective tests and should be followed. Reading and following directions is critical because students are often given a choice of questions to answer. Essay questions often contain specific directives indicating what to do and how to organize an answer. The following list includes common directives and their meanings:

Analyze: break into parts

Choose: select

Comment: similar to "Discuss" (below)

Compare: show similarities or similarities and differences

Contrast: show differences

Compare and contrast: show similarities and differences

Criticize: examine the pro and con positions and give your judgment

Define: give the meaning

Discuss: give as much information as you can

Evaluate: make a judgment and include the reasons that led to this judgment

Explain: give reasons

Illustrate: give examples

Interrelate: show relationships among

List: make the major points stand out clearly

Show: explain

State: give the information

Trace: show step-by-step development

You should also be alert to key words such as <u>after</u>, <u>before</u>, and <u>briefly</u> and phrases such as "two out of three," which specify and limit. By alerting yourself to these directives and key words, you can better organize your essay and avoid wasting time by providing information that is not required. The essay items in this study guide use several of these directives and key words. Note them as you use this guide and take essay exams.

Outline your answer before actually writing the essay to be certain you have included all the major points in logical order. If you run out of time, some instructors will give partial credit for material in outline form. Unless you are told differently, essays should be in paragraph form with a thesis sentence and a conclusion. Remember, many instructors also grade on spelling and punctuation, as well as on organization, so be certain your essay is grammatically correct. Your answer will also be enhanced if you use the correct terminology of your discipline.

I hope that you will find these techniques and strategies helpful. Learning and using proper study techniques will benefit you throughout your academic and professional career.

ESSAY QUESTIONS: UNDERSTANDING FORMAT

Margo Elliott

Students often complain that even though they studied their material diligently, they did not do well on their exams. This is particularly true in the case of essay questions. Students may feel that the instructor asked the "wrong" questions or stated them in a manner that was confusing or ambiguous.

More than likely, this is due in part to the way the information is learned and to how the information is understood. Memorizing facts, theories, and data is insufficient to answer questions that require anything more than simple knowledge. For facts to be useful, one must be able to understand and apply those facts, and this is what your instructors are testing.

Psychologists have suggested that there are different levels of comprehension and that these levels can be represented by a categorization scheme or taxonomy. One such system is Bloom's stage theory of cognitive development.[*] Bloom suggests six stages of understanding, each becoming successively more difficult and requiring a higher degree of comprehension.

Level I is the knowledge level. Knowledge questions require simple memorization and recall of various terms, theories, and research.

Level II is comprehension and involves the ability to understand nonliteral statements, such as examples, symbolism, and metaphors.

Level III requires the application of concepts or scientific terminology.

Level IV is analysis and uses the ability to recognize assumptions, to comprehend relationships, and to distinguish facts from hypotheses.

Levels V and VI involve synthesis and evaluation, respectively, and generally require a greater knowledge base of the discipline than is typically required in an introductory course. These cognitive abilities would more likely be used in upper-level and graduate work and will therefore not be discussed.

To further clarify, let's use a fictitious example.

Animal specialists have developed two methods you can use to teach your dog not to bark at the letter carrier. The first method (A) involves verbal punishment and reprimand when your dog (Sadie) barks at the letter carrier. The second method (B) uses reward and verbal praise when Sadie does not bark at the letter carrier. After months of training, the specialists concluded that although both methods worked, method B was more effective than method A.

Assume that you have read a text that presented a detailed account of the above research. Here are sample questions your instructor might ask that reflect the various levels. Notice how a greater degree of understanding is reflected in the increasing complexity of the questions.

---

[*]Bloom, Benjamin S., *Cognitive Domain* (New York: McKay, 1969).

Level I—Knowledge

1.   What is the dog's name?
2.   How many methods of training were discussed?
(These are simple recalls of concrete and specific information.)

Level II—Comprehension

1.   Explain the two methods of training.
2.   Give examples of appropriate reward and punishment for Sadie.
(Note the slightly higher level skills involved in answering these questions. Be sure to understand that "explain" does not mean "list.")

Level III—Application

1.   Using either method A or method B, devise a training program for a dog that barks at the letter carrier.
2.   How might these methods be used for training Sadie to shake hands?
(Note that these questions require applying an understanding of concepts and facts.)

Level IV—Analysis

1.   Explain why method B is a more successful training program than method A.
2.   Can these methods be generalized to child-rearing techniques? Be specific.
(Note that at this level individual thought is required.)

When preparing to study for exams, you will want to be alert to these various forms of content questions. It might also be helpful to ask your instructor if he or she would be willing to provide sample questions.

In the essay sections of each chapter, you will find questions that reflect the various levels. You may wish to refer back to this section when preparing for those questions.

C H A P T E R   1

GROUPS AND RELATIONSHIPS:
A SOCIOLOGICAL SAMPLER

Overview

Chapter 1 introduces the discipline of sociology: what it studies and how it differs from other social sciences. It opens with a discussion of the beginnings of sociology in the works of moral statisticians such as Quetelet, Guerry, and Morselli. Special attention is paid to Durkheim's work in the area of suicide. It then discusses the main focus of sociology—the group—and distinguishes among types of groups. It introduces both micro sociology and macro sociology. (The trend from micro to macro will be a major focus of this text.) It also discusses the scientific nature of sociology and highlights some of the challenges, drawbacks, and advantages of studying self-aware subjects. This chapter also offers examples of research techniques such as unobtrusive measures and validation research that are designed to minimize potential problems. Stanley Milgram's "small world" research and MacKay's study of parallel networks are highlighted as examples of "a closer view" of research. ("A Closer View" provides an in-depth look at the research process in action. This theme is carried throughout this text in individual chapters and provides a sense of sociology as an active discipline.) The chapter concludes with a brief discussion of the historical background of sociology and a discussion of the compatibility of the doctrine of free will and the social sciences.

Capsule Summary

Sociology began when moral statisticians such as Quetelet, Guerry, and Morselli began to study suicide rates in Europe. Emile Durkheim further elaborated upon these data and argued that high suicide rates reflect a weakness in the web of relationships among members of a society. Gradually, the study of moral statistics began to uncover the social causes of individual behavior, and sociology was born.

Sociology shares with the other social sciences an interest in human behavior. It differs from the other social sciences in its primary focus: the patterns and processes of human social relations. The study of group behavior is thus of major concern to sociologists. Indeed, sociology is often divided into micro sociology and macro sociology depending on the size of the group studied.

Groups may be large or small, but they all share the common characteristic of social relations among members. The smallest possible group is the dyad or group of two. Triads (groups of three) are of particular interest to sociologists because they often demonstrate the rules of transitivity and coalition formation. The patterns of social relations among members of a group are termed networks. Sociograms are used to study the structure of networks. Groups vary not only in size but also in the degree of intimacy shared by members. Groups in which the members share a good deal of intimacy are termed <u>primary groups</u>; those in which relationships are more impersonal are termed <u>secondary groups</u>.

Science is a method of discovery. <u>Science</u> uses <u>theories</u>, or very general statements, about how some portion of the world fits together and functions. Social scientists are in a unique position among scientists because their subjects are self-aware. <u>Unobtrusive measures</u> provide an interesting technique to test the accuracy of data. Although bias presents a potential problem in sociological research, the nature of scientific inquiry minimizes this bias. Science often seeks to challenge previous assumptions and theories.

<u>Mass society theorists</u> were concerned that <u>modernization</u> had led to a breakdown in social relationships among urban dwellers. (The theme of modernization will be dealt with extensively in later chapters.) Milgram's "small world" research and MacKay's study of <u>parallel networks</u> challenged many of these assumptions.

The origins of the social sciences can be traced to philosophy, but it was not until recently that research was conducted in the social sciences. Economist <u>Adam Smith</u> (1776) may be considered the first real social scientist. <u>Auguste Comte</u> used the term <u>sociology</u> in the 1830s, and early European sociologists included <u>Spenser</u>, <u>Tönnies</u>, and <u>Durkheim</u>. <u>Albion Small</u> and <u>W. E. B. DuBois</u> were important in the development of sociology in the <u>United States</u>.

The doctrine of <u>free will</u> is compatible with the social sciences even though on the surface it may appear contradictory. All social science theories assume that humans possess the ability to reason and make choices. These choices are, however, predictable because people will choose what they find rewarding given their circumstances, information, and preferences.

Key Concepts

You should be able to explain the concepts in the following list and be able to cite several examples of each. The page number after each indicates where it is introduced.

Moral statistics 3	Social network 13
Sociology 5	Sociogram 13
Social sciences 5	Internal faction (clique) 13
Unit of analysis 6	Primary group 15
Micro sociology 9	Secondary group 16
Macro sociology 9	Theory 17
Group 9	Research 18
Aggregate 12	Unobtrusive measure 18
Dyad 12	Validation research 20
Triad 12	Self-report studies 20
Transitivity 13	Chains of attachments 21
Intransitive triad 13	Parallel networks 25
Coalition 13	Religious determinism (fatalism) 28
Power 13	Free will 28

Key Research Studies

You should be familiar with both the methodology and the results of the research studies cited here.

Durkheim: *Suicide*—study of suicide rates based on Morselli's data 4

Hypothetical study using units of analysis 6

Litwak and Messeri: primary group membership and "natural" causes of death 15

Bainbridge and Stark: use of unobtrusive measures in studying geographical patterns in metaphysical beliefs and practices 18

Hirschi: delinquency study using validation research 21

Milgram: "small world" research networks 24

J. Ross MacKay: parallel social networks in Canada 25

Key Theories

You should be prepared to explain the assumptions of these theories and, when applicable, to cite related research findings.

Micro sociology

Macro sociology

Free will

In addition to these key theories this chapter discusses several key figures in the early development of sociology. You should be able to associate each person with his contribution.

Andre Michel Guerry: cofounder of moral statistics 3

Adolphe Quetelet: cofounder of moral statistics 3

Henry Morselli: gathered statistics on suicide 4

Charles H. Cooley: coined the term *primary group* 16

Adam Smith: economist and first social scientist 27

Auguste Comte: coined the term *sociology* 27

Herbert Spenser: published *Principles of Sociology* 27

Ferdinand Tönnies: published *Gemeinschaft und Gesellschaft* 27

Albion Small: founded first sociology department in the United States at the University of Chicago 27

W. E. B. DuBois: created a sociological laboratory and directed the Atlanta University Conferences 27

Completion

Although a statement may show only one fill-in blank, some blanks may require two or more words for completion.

1. The cofounders of moral statistics were _____ and _____.

2. Durkheim argued that _____ suicide rates reflect weaknesses in the web of relationships among members of society.

3. The "things" on which a set of research observations are based are termed _____.

4. _____ sociologists tend to focus on large groups and whole societies.

5. The smallest social group is a(n) _____.

6. "Any friend of yours is a friend of mine" demonstrates the rule of _____.

7. The ability to get one's way over the opposition of others is termed _____.

8. The patterns of relationships among members of a group are often called _____.

9. _____ groups are characterized by great intimacy among members.

10. _____ measures obtain information without disturbing the objects of the research.

11. Comparison of results when different measures are used is one way to assess the _____ of sociological data.

12. Milgram and MacKay both studied the existence of _____.

13. MacKay discovered that language barriers within Canada are _____ than are national boundaries.

14. The essence of the scientific method is _____.

15. The proper approach to research is to try to _____ those things that the researcher actually believes to be true.

16. Personal bias is possibly a more serious problem in the _____ than in the natural or physical sciences.

17. Scientific explanations must take the form of theories, and these must be the object of testing by _____.

18. The term *sociology* was first suggested by _____.

19. _____ created a sociological laboratory and directed the Atlanta University Conferences.

20. The doctrine of _____ argues that humans possess the capacity for choosing among alternatives and, therefore, can be held responsible for the choices they make.

Multiple-Choice

1. Nineteenth-century statistics comparing suicide rates from one nation to another showed that:
   a. the rates were extremely stable year to year.
   b. the rates varied little from one nation to another.
   c. during the nineteenth century the rates declined sharply.
   d. all of the above
   e. none of the above

2. Emile Durkheim argued that:
   a. traditional rural societies were deficient in the kinds of warm interpersonal relationships that are typical of modern societies.
   b. high suicide rates reflect a weakness in an individual's personality.
   c. high suicide rates reflect weaknesses in the web of relationships among members of society.
   d. a and c
   e. a and b

3. The primary subject of sociology is:
   a. the individual.
   b. the group.
   c. preliterate societies.
   d. illegal behavior.
   e. political organization.

4. Units of analysis include:
   a. individuals.
   b. groups.
   c. states.
   d. a and b above
   e. all of the above

5. The *main* difference between micro sociology and macro sociology is the:
   a. size of the group studied.
   b. research method used.
   c. training of the researcher.
   d. degree of industrialization achieved by the group studied.
   e. presence of self-aware subjects in micro sociology.

6. The social sciences include:
   a. anthropology.
   b. political science.
   c. sociology.
   d. a and c
   e. all of the above

7. Intransitive triads:
   a. are demonstrated by the statement, "Any friend of yours is a friend of mine."
   b. are unstable and usually break up.
   c. may lead to coalition formation—two against one.
   d. all of the above
   e. b and c

8. A triad is:
   a. a group of three.
   b. the smallest sociological group.
   c. always intransitive and unstable.
   d. all of the above
   e. a and c

9. Primary group members:
   a. share a good deal of intimacy with one another.
   b. gain much of their self-esteem from their groups.
   c. often refer to themselves as "we."
   d. a and c
   e. all of the above

10. Which of the following would most likely *not* be considered a primary group?
    a. a political party
    b. a group of intimate friends
    c. a family
    d. b and c
    e. none of the above

11. When researchers test data against some independent standard of accuracy, they are:
    a. using an unobtrusive measure.
    b. conducting validation research.
    c. using self-reports.
    d. a and c
    e. none of the above

12. In his study of delinquency, Travis Hirschi:
    a. used self-reports.
    b. conducted validation research.
    c. used both interviews and questionnaires.
    d. a and b
    e. all of the above

13. Milgram's "small world" research:
    a. lent considerable support to mass society theories.
    b. found that most of the letters did not reach their designated receiver.
    c. discovered that people throughout the country were united by "chains of attachment."
    d. discovered a very strong interest in astrology in the Far West.
    e. a and b

14. The researcher who asked auto mechanics to note the position of the radio dials in cars they serviced to ascertain what stations people listened to while driving used:
    a. an unobtrusive measure.
    b. validation research.
    c. self-reported behavior studies.
    d. b and c
    e. all of the above

15. The term *sociology* was first suggested by:
    a. Adolphe Quetelet.
    b. Auguste Comte.
    c. Emile Durkheim.
    d. Charles H. Cooley.
    e. Albion Small.

16. Which of the following early sociologists is *not* correctly paired with his contribution?
    a. Ferdinand Tönnies: published *Gemeinschaft und Gesellschaft*
    b. Emile Durkheim: suggested the term *sociology*
    c. Albion Small: founded the first sociology department in North America
    d. a and b
    e. b and c

17. Recent research has shown that:
    a. early sociologists such as Durkheim were correct in their assumption that modernization substantially increases social isolation.
    b. membership in a primary group greatly reduces many "natural" causes of death.
    c. there are few, if any, primary groups in our large, impersonal cities.
    d. all of the above
    e. none of the above

18. Reasons for the existence of parallel networks include:
    a. racism.
    b. language.
    c. distance.
    d. a and b
    e. all of the above

19. Which of the following statements is/are true?
    a. The essence of the scientific method is systematic skepticism.
    b. The purpose of scientific research is to test what we believe about the world.
    c. Because bias is a problem in sociological research, sociologists must completely rid themselves of personal biases before beginning research.
    d. a and b
    e. all of the above

20. The doctrine of free will:
    a. is incompatible with the social sciences.
    b. assumes that humans have the ability to reason and to make choices.
    c. assumes that people will seek those things they find rewarding and will avoid those they find unrewarding.
    d. b and c
    e. all of the above

Essay

In each chapter, questions 1 and 2 are designed to reflect the cognitive levels described in the Introduction. In each case, the general topic in question is divided into three smaller questions, each requiring a different level of understanding. Some questions tap levels of knowledge, comprehension, and application, while others tap comprehension, application, and analyses. The level is indicated in parentheses. You might wish to refer to the Introduction for a more detailed explanation of the cognitive level in question. In addition, each set of essay questions will contain one question (question 6) which will suggest ways in which you can apply some of the information in this chapter to everyday life. While there will be no one answer to this question, it will encourage you to apply the theories and research findings to practical situations you may encounter. Admittedly, some of these essays are fairly long and difficult, but by using them for practice you should be able to test yourself on how well you have mastered the material in question.

1.   A.   Name some research techniques used to study self-aware subjects. (knowledge)
          B.   Explain these research techniques. (comprehension)
          C.   Design a simple study that uses one of the above techniques. (application)

2.   A.   Name some assumptions of mass society theorists. (knowledge)
          B.   Explain Milgram's research. (comprehension)
          C.   Show how Milgram's results did or did not support mass society theory. (analysis)

3.   Trace the development of the social sciences from their origins in philosophy to the present.

4.   Discuss the principles of transitivity and coalition formation. Explain how the study of networks is relevant to both micro and macro sociology.

5.   Explain the following statement: "It is only because people's choices are predictable that it is possible to claim that they have free will."

6.   Suppose you plan to transfer to another university where you do not know any other students. How could you use social networks to arrange contacts with students at this university during the transfer process?

Answers

Completion

1.	Adolphe Quetelet, Andre Guerry	11.	validity
2.	high	12.	social networks
3.	units of analysis	13.	less powerful
4.	Macro	14.	systematic skepticism
5.	dyad	15.	disprove
6.	transitivity	16.	social sciences
7.	power	17.	systematic research
8.	social networks	18.	Auguste Comte
9.	Primary	19.	W. E. B. DuBois
10.	Unobtrusive	20.	free will

Multiple-Choice

1.	a	11.	b
2.	c	12.	e
3.	b	13.	c
4.	e	14.	a
5.	a	15.	b
6.	e	16.	b
7.	e	17.	b
8.	a	18.	e
9.	e	19.	d
10.	a	20.	d

CHAPTER 2

CONCEPTS FOR SOCIAL AND CULTURAL THEORIES

Overview

Part I opens with a discussion of the eight steps in theory construction and theory testing and the importance of concepts and theories in science. Chapter 2 begins with a discussion of two important concepts in sociology: society and culture. Patterns of intergroup relations are discussed; social stratification is introduced; and the concepts of class, mobility, achieved status, and ascribed status are explained. Cultural components such as norms, values, and roles also are discussed. Cultural and social theories of assimilation are then illustrated through the works of Zborowski and Herzog, Covello, Steinberg, Perlmann, and Greeley. The chapter closes with a look at the importance of reference groups.

Capsule Summary

Scientific theories exist to explain why. Theories are general in that they apply to all instances and make predictions that can be checked out. The scientific process includes both theory construction and theory testing. A sociologist engaged in theory construction and testing would first begin by wondering and then proceed to conceptualize, theorize, operationalize, hypothesize, observe, analyze, and lastly assess. Concepts are used in science to classify things that are alike. They are abstract and serve as the building blocks for theories, which are statements that say why and how several concepts are related. Although theories are general, hypotheses tell the implications of predictions in specific situations.

   Society and culture are essential concepts in sociology. A society is a group of people characterized by social relationships, relative self-sufficiency and independence, duration over time, a physical location, and a common culture. Culture, on the other hand, is the sum total of human creation—intellectual, technical, artistic, physical, and moral. It is the complex pattern of living that directs social life, the things each new generation must learn and to which they eventually add. Norms, values, and roles are important components of culture. Norms refer to rules or guidelines for behavior, while values serve as standards for assessing the desirability or undesirability of something; norms and values are often related. Roles refer to collections of norms associated with particular positions in society. Cultures differ not only in their actual norms, values, and roles, but also in the relative importance attached to them.

   Social stratification refers to the unequal distribution of rewards among members of a society. Societies are "layered," and these layers constitute classes— groups of people sharing a similar position (or status) in a society. Movement within a stratification system is termed mobility, and the direction it takes may be upward or downward. Position (status) within the stratification system may be based on either achievement or ascription.

   When members of one culture migrate to another, different patterns of intergroup relations may occur. If the immigrant group gives up its old culture and totally adopts the new, then <u>assimilation</u> has occurred. If members of the new culture resist full acceptance of the immigrant group, then <u>prejudice</u> and <u>discrimination</u> against the immigrant group may result. In such cases, the immigrant group may become a <u>subordinate group</u> in the new culture. In other cases, <u>accommodation</u> may occur, which results in <u>cultural pluralism</u>. Often the immigrant group becomes a <u>subculture</u> within the new culture.

During the nineteenth and twentieth centuries, many immigrant groups were the objects of <u>prejudice</u> and <u>discrimination</u> when they arrived in the United States and Canada. The experiences of both Jewish and Italian immigrants in the United States provide interesting examples of cultural differences. <u>Zborowski</u> and <u>Herzog</u> studied <u>shtetl</u> life in Europe to gain information on the cultural background of Jewish immigrants. They concluded that this background was a major factor in facilitating <u>upward mobility</u> after immigration. <u>Steinberg</u> further investigated this area and found that the <u>occupational background</u> of Jewish immigrants also facilitated upward mobility, thus implying a more social rather than cultural explanation. <u>Perlmann</u> was able to synthesize both the social and cultural explanations in his study of schooling and occupational achievements among ethnic groups in Rhode Island. In his study of Italian immigrants, Covello found that their cultural background initially served as a deterrent to upward mobility. <u>Greeley</u>'s study of the <u>persistence of Italian culture</u> found that Italian family <u>values</u> have not changed, but the <u>norms</u> by which families fulfill their values have changed. Their <u>reference group</u> identification probably was also a factor because most Italian immigrants considered their <u>reference group</u> to be those who remained in Italy. Thus, the research studies indicate that differences in cultural background may partly explain the differing rates of initial upward mobility between Jewish and Italian immigrants. Recent data indicate that prejudice against Jews (<u>anti-Semitism</u>) and Italians in the United States and Canada seems to have diminished markedly as a result of <u>assimilation</u> (particularly intermarriage) and <u>accommodation</u>.

Key Concepts

You should be able to explain the concepts listed here as well as cite several examples of each.

Concept 34	Prejudice 45
Hypothesis 35	Discrimination 45
Society 41	Value 45
Culture 43	Norm 45
Technology 43	Role 45
Stratification 44	Assimilation 47
Class 44	Accommodation 48
Status 44	Cultural pluralism 48
Upward mobility 44	Subculture 48
Downward mobility 44	Anti-Semitism 49
Achieved status 44	Ghetto 55
Ascribed status 44	Reference group 66

Key Research Studies

You should be familiar with both the methodology and the results of these researchers and their studies.

          Glock and Stark, Stark, and others: anti-Semitism in the United States 53
          Zborowski and Herzog: cultural backgrounds of Jewish immigrants 55
          Covello: cultural background of Italian immigrants 57
          Steinberg: cultural and occupational backgrounds of Jewish immigrants 59
          Perlmann: schooling and occupational achievements among ethnic groups in
               Rhode Island 63
          Greeley: the persistence of Italian culture 70

Key Theories

          Eight steps in theory construction and theory testing
          Cultural theory
          Social theory

Completion

1. _____ are abstractions that are used to classify sets of things that are alike.

2. A(n) _____ is a group of people who are united by special relations, and it is relatively self-sufficient and independent.

3. _____ is the sum total of human creations—intellectual, technical, physical, and moral.

4. _____ are statements that say why and how several concepts are related.

5. A(n) _____ tells us the implications of a prediction in the specific situation we observe.

6. Stratification is the _____ distribution of rewards among members of a society.

7. When a lawyer's daughter becomes a factory worker, we can say that she has experienced _____ mobility.

8. _____ statuses are derived from inheritance, whereas _____ statuses are derived from individual merit.

9. _____ refers to negative attitudes toward a group, while _____ refers to actions taken against a group.

10. Norms are _____ that govern behavior.

11. The _____ of a culture identify its ideals.

12. A(n) _____ is a collection of norms associated with a particular position in a society.

13. The process of exchanging one culture for another is termed _____.

14. _____ occurs when accommodation results in the continued existence of several distinctive cultures within a society.

15. A distinctive set of beliefs, morals, customs, and the like that is developed or maintained by a group within a larger society is called a(n) _____.

16. Prejudice against Jews is termed _____.

17. Marrying someone of another ethnic background is called _____.

18. Zborowski and Herzog found that the cultural background of Jews stressed _____ as an important value.

19. Covello found that the cultural background of Italian immigrants did not serve to foster their achievement of _____.

20. Groups that individuals identify with and whose norms and values serve as their basis for self-judgment are termed _____.

Multiple-Choice

1. Which of the following statements is/are true?
   a. Concepts are names used to identify some set or class of things that are said to be alike.
   b. Scientific concepts are concrete; they identify things, not ideas.
   c. Concepts are the building blocks of any science.
   d. a and c
   e. all of the above

2. Scientific theories:
   a. are specific in nature.
   b. must include or imply some conclusions that can be empirically verified by direct physical observation.
   c. are more specific than hypotheses.
   d. a and b
   e. all of the above

3. Which of the following are the *last* three steps in the scientific process?
   a. observation, analysis, and assessment
   b. operationalization, observation, and analysis
   c. observation, conceptualization, and analysis
   d. conceptualization, analysis, and assessment
   e. none of the above

4. Characteristics of societies include:
   a. a definite physical location.
   b. relative self-sufficiency and independence.
   c. existence over time.
   d. a and c
   e. all of the above

5. Culture is:
   a. the sum total of human creation.
   b. often synonymous with nation.
   c. unchanged by each new generation.
   d. a and b above
   e. all of the above

6. When a factory worker's child becomes a physician, we say that he or she has experienced:
    a. upward mobility.
    b. downward mobility.
    c. circular mobility.
    d. horizontal mobility.
    e. none of the above

7. Ascribed statuses may be based on:
    a. family background.
    b. individual merit.
    c. genetic inheritance.
    d. a and c
    e. all of the above

8. The caste system in India:
    a. is an extreme example of a stratification system based on achieved status.
    b. is based on an individual's ability or merit.
    c. is an extreme example of a stratification system based on ascribed status.
    d. a and b
    e. all of the above

9. Rules governing behavior are termed:
    a. values.
    b. beliefs.
    c. norms.
    d. roles.
    e. none of the above

10. Which of the following statements is/are true?
    a. The values of a culture identify its ideals.
    b. Norms are quite general, whereas values are specific.
    c. Values justify the norms.
    d. a and c
    e. all of the above

11. _____ refers to negative attitudes toward a group, while _____ refers to negative actions against a group.
    a. Prejudice, discrimination
    b. Discrimination, prejudice
    c. Subordination, discrimination
    d. Prejudice, assimilation

12. Which of the following statements is/are true?
    a. Social life is structured by roles.
    b. Cultures differ in their evaluation of various roles.
    c. Some roles are thought to be more demanding than others.
    d. b and c
    e. all of the above

13. The process of exchanging one culture for another is termed:
    a. assimilation.
    b. accommodation.
    c. cultural pluralism.
    d. discrimination.
    e. cultural exchange.

14. The existence of different religions side by side in the United States today is an example of:
    a. cultural pluralism.
    b. assimilation.
    c. subordinate groups.
    d. discrimination.
    e. cultural lag.

15. In his study of educational and occupational achievement among ethnic groups, Perlmann found that:
    a. Jewish sons had lower high school graduation rates than did their Yankee or Italian counterparts.
    b. social theory alone is adequate to explain the full range of Italian-Jewish differences in educational achievement.
    c. cultural theory alone is adequate to explain the full range of differences in educational achievement.
    d. a synthesis of cultural and social theories is needed to explain the differences among the groups.
    e. innate differences partly explain the differences among the groups.

16. According to Covello, immigrants from southern Italy brought with them to the United States:
    a. a "cult of scholarship" and emphasis on learning.
    b. the belief that school was harmful and a threat to family loyalty.
    c. extensive training in the professional and middle-class occupations.
    d. a and c
    e. none of the above

17.   Steinberg found that many of the immigrant Jews:
        a.   had been farmers and, hence, could only find employment in unskilled
             occupations.
        b.   were trained in skilled and professional occupations.
        c.   were single men who planned to return to their families in Europe.
        d.   a and c
        e.   none of the above

18.   In his study of the persistence of Italian culture, Greeley found:
        a.   both Italian family values and the norms by which they fulfill their values
             have changed.
        b.   Italian family values have not changed but the norms by which they fulfill
             their values have changed.
        c.   neither Italian family values nor the norms by which they fulfill their
             values have changed.
        d.   Italian family solidarity continues to impede success.
        e.   none of the above

19.   Studies by Stark and others on anti-Semitism found that by the middle and
        late 1960s:
        a.   prejudice against Jews was increasing.
        b.   prejudice against Jews had declined greatly.
        c.   prejudice against Italians had declined.
        d.   prejudice against Italians had increased.
        e.   none of the above

20.   Reference groups:
        a.   are groups with which individuals identify.
        b.   are the people whose approval counts most with us.
        c.   must be present to influence a person's behavior.
        d.   a and b
        e.   all of the above

Essay

1.   A.   Describe the research findings of Zborowski and Herzog, Steinberg,
             Perlmann, and Greeley. (knowledge)
        B.   Do you feel the cultural backgrounds of other minorities have
             influenced their mobility? Give specific examples. (comprehension)
        C.   Contrast the cultural backgrounds of Jewish and Italian immigrants
             and show how these backgrounds influenced their mobility.
             (analysis)

2.   A.   Briefly explain the terms <u>norm</u>, <u>value</u>, and <u>role</u>. (knowledge)
       B.   Give two examples of the above terms. (comprehension)
       C.   Interrelate these concepts using specific examples. (analysis)

3.   Discuss some of the characteristics and uses of scientific concepts.

4.   Distinguish between society and culture and discuss the characteristics of each.

5.   Discuss patterns of intergroup relations that might occur when immigrant groups arrive in a new culture. Give examples from U.S. history.

6.   You discover that your roommate is from a different ethnic or cultural background. You have noticed that he or she has some very different attitudes toward study and academic achievement, family relationships and ties, and educational aspirations than those you hold. How might some of the material in this chapter, especially the research findings of Herzog, Covello, etc., lend insight into the origin of these differences? If you both feel comfortable you may wish to share information about your backgrounds to better understand these differences and to enhance your mutual appreciation for cultural and ethnic diversity.

Answers

Completion

1.	Concepts	11.	values
2.	society	12.	role
3.	Culture	13.	assimilation
4.	Theories	14.	Cultural pluralism
5.	hypothesis	15.	subculture
6.	unequal	16.	anti-Semitism
7.	downward	17.	intermarriage
8.	Ascribed, achieved	18.	education
9.	Prejudice, discrimination	19.	upward mobility
10.	rules	20.	reference groups

Multiple-Choice

1.	d	11.	a
2.	b	12.	e
3.	a	13.	a
4.	e	14.	a
5.	a	15.	d
6.	a	16.	b
7.	d	17.	b
8.	c	18.	b
9.	c	19.	b
10.	d	20.	d

MICRO SOCIOLOGY: TESTING INTERACTION THEORIES

Overview

Chapter 3 opens with a discussion of scientific theories of micro sociology such as rational choice theories, interaction theories, and symbolic interaction. Cooley and Mead's works on socialization and the development of the self are explained. The importance of attachments is introduced and discussed. (This theme will be developed throughout the text.) The chapter then discusses research, with emphasis on the importance of ascertaining causation. Techniques for determining causation are described. Ofshe's experimental study of attachments and conformity and the author's study with Lofland on attachments and conversion are described in "A Closer View." A discussion of replication concludes the chapter.

Capsule Summary

Rational choice theory, exchange theory, and symbolic interaction are examples of micro theories in sociology. Micro theories assume that people make choices based on rewards and costs. Rational choice theories assume that because rewards are typically obtained from others, people engage in exchange relationships and over time tend to establish stable exchange partnerships with others. Attachments emerge from such relationships, and norms function to ensure some basis for predictability. Symbolic interaction focuses on the importance of symbols, which stand for or indicate other things. Symbols are essential for human communication. Cooley and Mead, founders of symbolic interaction, focused on the processes of socialization and the development of the self.

Research is the process of making systematic observations. Much research focuses on testing specific hypotheses. Research tries to establish causation by establishing the presence of the three criteria of causation: correlation, time order, and nonspuriousness. Correlation can be established if it can be shown that two things vary or change in unison. Time order can be demonstrated if it can be shown that the cause (often termed the independent variable) precedes the effect (often termed the dependent variable). Nonspuriousness can be established if it can be shown that the effect was not produced by something else. Typically, research studies are replicated by others to determine whether the same results are consistently obtained.

Sociological research may be either experimental or nonexperimental. Nonexperimental research is probably more common because many of the problems typically studied by sociologists do not lend themselves well to laboratory research. Experiments offer greater control over the subjects. Nonexperimental research findings are more subject to risk. Ofshe's study of attachments and conformity is an example of experimental research. He was able to demonstrate causation by establishing the three criteria. He used controls and randomization to rule out the possibility of spuriousness and employed a test of significance to rule out chance. Stark and Lofland's study of attachments and conversion is an interesting example of nonexperimental research. They too were able to establish <u>correlation</u> and <u>time order</u>; but, as is often the case with <u>observational studies</u>, <u>nonspuriousness</u> was more difficult to ascertain.

Key Concepts

You should be able to explain the concepts listed here and be able to cite several examples of each.

Micro sociology 75	Attachment 83
Rational choice (self-interest) proposition 78	Causation 85
	Correlation (positive and negative) 86
Altruism 79	Spuriousness 86
Reward 80	Variable 88
Cost 80	Independent variable 88
Exchange relations 80	Dependent variable 88
Goods 80	Experimental control 89
Social interaction 80	Randomization 89
Symbol 80	Test of significance 89
Self 81	Nonexperimental research 90
Socialization 81	
"Looking glass self" 81	Field observation 91
Mind (Mead) 82	Replication 94
"Taking the role of the other" 82	

Key Research Studies

You should be familiar with both the methodology and the results of the research studies cited here.

Dion: influence of appearance on perception of children's misbehavior 76

Ofshe: study of attachments and conformity among college students (experimental study) 87

Stark, Lofland, and others: nonexperimental research on conformity and conversion 90

Key Theories

You should be prepared to explain the assumptions of these theories and, when applicable, to cite related research findings.

        Micro sociology theories (in general)
        Rational choice theories
        Symbolic interactionism (including Cooley and Mead's work)
        Exchange theory

Completion

1. When a theory pertains to the behavior of individuals or small groups, it is in the realm of _____ sociology.

2. All micro theories in the social sciences assert that _____ is the most basic aspect of human behavior.

3. The more formal variety of interaction theories in sociology are referred to as _____ theories, whereas the less formal, older variety are referred to as _____ theories.

4. Social science proceeds on the principle that, given our options and preferences, we choose to do that which we can expect to be most _____.

5. Psychologists believe behavior is shaped by _____.

6. Humans seek what they perceive to be _____ and avoid what they perceive to be_____.

7. Unselfish behavior to benefit others is termed _____.

8. _____ is the process by which we influence one another.

9. _____ are things that stand for or indicate other things and are of primary importance to the theory of _____.

10. Both Cooley and Mead concluded that each person's sense of self is _____.

11. Mead used the concept of _____ to identify understanding of symbols and _____ to identify our learned understanding of the responses of others.

12. Mead termed the ability to put ourselves in another's place the ability to _____.

13. Over time, people tend to establish stable _____ partnerships.

14. A stable and persistent pattern of interaction between two people is termed a(n) _____.

15. _____ is the process of making systematic observations.

16. _____ are specific predictions about the empirical or observable world.

17. The three criteria of causation are _____, _____, and _____.

18. The independent variable indicates a(n) _____, while the dependent variable indicates a(n) _____.

19. Ofshe found a significant correlation between _____ and conformity among students.

20. Stark and Lofland's study of _____ and conversion is an example of _____ research.

_____

Multiple-Choice

1.   Which of the following statements is/are true?
        a. All micro theories in the social sciences assert that choice is the most basic aspect of human behavior.
        b. Sociologists differ from economists in that economists greatly expand the concepts of rewards and costs.
        c. Exchange theory is an example of a macro sociology theory.
        d. a and b
        e. none of the above

2.   Rational choice theories include:
        a. symbolic interaction theory.
        b. conflict theory.
        c. exchange theory.
        d. a and c
        e. all of the above

3.   Micro sociology departs from the other micro social science theories in that micro sociology:
        a. expands the concepts of reward and cost.
        b. recognizes that much of what we want can only be gotten from others.
        c. defines goods as the whole range of rewards that people seek.
        d. a and b
        e. all of the above

4.   Micro sociology consists primarily of the study of:
       a.   individuals.
       b.   face-to-face interaction in small groups.
       c.   the regularities and patterns that arise out of interaction and exchanges.
       d.   b and c
       e.   none of the above

5.   Symbols:
       a.   are things that stand for or indicate other things.
       b.   are important for human communication.
       c.   are part of our genetic makeup.
       d.   a and b
       e.   all of the above

6.   The "looking glass self" is associated with:
       a.   Charles H. Cooley.
       b.   George H. Mead.
       c.   Richard Ofshe.
       d.   John Lofland.
       e.   George Homans.

7.   Mead considered the mind to be:
       a.   our learned understanding of the responses of others to our conduct.
       b.   our understanding of symbols.
       c.   an innate component of the brain.
       d.   all of the above
       e.   none of the above

8.   Mead argued that a child cannot take an effective part in most games until he
       or she:
       a.   can take the role of the other.
       b.   has developed a superego.
       c.   has developed a sense of self.
       d.   a and c
       e.   all of the above

9.   Stable exchange partnerships:
       a.   are of special importance to us.
       b.   are restricted to goods and services.
       c.   often lead to the forming of attachments.
       d.   a and c
       e.   all of the above

10. Specific predictions about the empirical or observable world are termed:
    a. theories.
    b. variables.
    c. controls.
    d. hypotheses.
    e. norms.

11. Criteria for establishing causality include:
    a. replication, time order, and correlation.
    b. correlation, time order, and spuriousness.
    c. correlation, nonspuriousness, and time order.
    d. correlation, spuriousness, and time order.
    e. correlation, time order, and replication.

12. _____ occurs when two factors appear correlated, with one seeming to cause the other, when the correlation actually is caused by a third, unnoticed factor.
    a. Nonspuriousness
    b. Spuriousness
    c. A negative correlation
    d. An inverse correlation
    e. none of the above

13. A positive correlation exists when:
    a. both factors decline.
    b. both factors increase.
    c. one factor rises while the other declines.
    d. no relationship exists between the factors.
    e. a and b

14. The term _____ variable is used to indicate a cause.
    a. dependent
    b. independent
    c. intervening
    d. spurious
    e. none of the above

15. To determine the criteria of nonspuriousness, Ofshe used:
    a. replication.
    b. randomization.
    c. controls.
    d. b and c
    e. all of the above

16. In determining significance, both _____ and _____ are taken into account.
    a. number of subjects, size of the correlation
    b. number of subjects, time order of the experiment
    c. size of the correlation, time order of the experiment
    d. number of subjects, ages of the subjects
    e. none of the above

17. Stark and Lofland's study is an example of:
    a. field observation research.
    b. the experimental method.
    c. replication research.
    d. secondary research.
    e. all of the above

18. Stark and Lofland concluded that the primary basis for conversion to the Unification Church was:
    a. ideology.
    b. brainwashing.
    c. social class membership.
    d. attachments.
    e. a and b

19. Replication studies of religious conversion have found that:
    a. it is impossible to replicate nonexperimental research.
    b. there is little support for the findings of Stark and Lofland's earlier study.
    c. ideology rather than attachments plays the major role in determining conversion.
    d. attachments play the major role in determining conversion.
    e. b and c

20. _____ is/are the most common form of symbolic interaction.
    a. Gestures
    b. Conversation
    c. Body language
    d. Reading
    e. Writing

Essay

1.   A.   Name the three criteria of causation. (knowledge)
          B.   Explain the three criteria of causation. (comprehension)
          C.   Show how Ofshe was able to determine causation in his experiment. (analysis)

2.   A.   Define experimental and nonexperimental research. (knowledge)
          B.   Design a study using either experimental or nonexperimental research. (application)
          C.   Using examples from the text, *compare and contrast* experimental and nonexperimental research. (analysis)

3.   Distinguish between a theory and a hypothesis. Explain the following statement: "Researchers do not contribute to scientific progress by seeking evidence that will support theories but by doing their utmost to disprove them."

4.   The premise that people make choices is central to micro theories in social science. Explain how this premise is incorporated by micro sociology.

5.   Briefly explain symbolic interaction. Explain the theories of Cooley and Mead on the development of the self.

6.   This chapter dealt with the importance of attachments and focused on research by Ofshe on the theory of loyalty. Assume you are the leader of a group; perhaps a student organization, a civic group, or a committee in a workplace. You have noticed that when controversial issues are raised certain members of the group voice the same opinions and typically vote as a block. This frequently causes other members to fail to voice their opinions and, as a result, often feel outnumbered and tend to lose interest or even drop out. How can you use your knowledge of loyalty theory to explain this and devise methods which would increase the participation of the other members to assure that decisions will be made in a more democratic manner?

Answers

Completion

1.	micro	11.	mind, self
2.	choice	12.	take the role of the other
3.	exchange, symbolic interaction	13.	exchange
4.	rewarding	14.	attachment
5.	reinforcement	15.	Research
6.	rewards, costs	16.	Hypotheses
7.	altruism	17.	correlation, time order,
8.	Social interaction		nonspuriousness
9.	Symbols, symbolic	18.	cause, effect
	interactionism	19.	attachments
10.	socially created	20.	attachments, nonexperimental

Multiple-Choice

1.	a	11.	c
2.	d	12.	b
3.	e	13.	e
4.	d	14.	b
5.	d	15.	d
6.	a	16.	a
7.	b	17.	a
8.	d	18.	d
9.	d	19.	d
10.	d	20.	b

CHAPTER 4

MACRO SOCIOLOGY: TESTING STRUCTURAL THEORIES

Overview

Chapter 4 opens with examples of research on bystander apathy and the relationship between religion and delinquency that illustrate the relationship between micro and macro sociology. The importance of attachments (a major theme of this text) is again emphasized. It then describes survey research and continues with a discussion of the elements of systems and the subject matter of macro sociology. The chapter discusses in-depth the three theories of macro sociology: functionalism, social evolution, and conflict. It closes with "A Closer View," Jeffery Paige's cross-cultural study of family structures and political conflict. At the end of this chapter is a special topic devoted to correlation and sampling.

Capsule Summary

Macro sociology is concerned with the study of social structures: groups, institutions, organizations, and societies. Micro and macro sociology are related, and often research in one area may overlap into the other. In their studies of delinquency and religious commitment, for example, Hirschi and Stark and others began by examining areas typically associated with micro sociology but discovered that it was necessary instead to focus on areas in the realm of macro sociology.

Theories of macro sociology attempt to explain the existence of social structures and their origins, differences, and interrelationships, as well as the interplay between the individual and the social structure. Macro sociology views societies as social systems characterized by separate structures, interdependency, and equilibrium. Macro sociology often studies institutions and classes. Institutions are clusters of roles, groups, organizations, customs, and activities that meet the basic needs of a society. Every society has at least five basic institutions: the family, the economy, religion, the political order, and education. Classes are groups of people who share a similar position in a society's stratification system. Macro sociology thus focuses on institutions and classes not only as separate structures but also as interrelated parts striving toward equilibrium.

Macro sociological theories include functionalism, social evolution, and conflict theory. Each focuses on social structures, and yet each emphasizes different aspects of these structures and makes different assumptions about them.

Functionalism is concerned with functions, or the part that each element of the system contributes to the whole. Functional theories have three components: They identify and explain (1) an aspect of the system, (2) that aspect's existence in terms of how it preserves another part from disruption, and (3) the source of potential disruption. Functionalism also focuses on functional alternatives and dysfunctions.

Social evolution theories focus on the development of social structures over time and how they adapt to their physical and social environments. They postulate that those societies that have developed the best-adapted structures tended to grow and become more powerful and complex. They focus on all societies and do not make value judgments regarding the direction of change.

Conflict theory considers conflicts within a structure that arise from the differing interests of competitive groups. It emphasizes how the structure may be shaped by the interests of these groups, especially the more powerful ones who serve their needs at the expense of the less powerful. Marx argued that the social structures are created by the ruling class. Weber expanded the concept and emphasized the importance of status groups.

Macro sociology often uses survey research. Samples are drawn, and questionnaires or interviews are administered. Attention is paid to ensuring that the criteria of causality are met. (The studies of Paige, Stark, and others illustrate this.) Although macro sociology typically focuses on groups, institutions, and organizations, occasionally whole societies are compared and contrasted. Paige's research on kinship structure and political conflict used such a cross-cultural comparison and is an example of research in macro sociology.

Key Concepts

You should be prepared to explain the concepts listed here and to cite several examples of each.

Macro sociology 97	Nuclear family 108
Social structure 98	Extended family 108
Survey research 100	Functional alternative 109
Sample 100	Dysfunction 110
System 105	Social evolution 110
Institution 105	The ruling class 111
Class 106	Status group 111
Stratification 106	Kinship 115
Equilibrium 107	Patrilocal residence 115
Functionalism 108	Matrilocal residence 116

The following concepts are contained in Special Topic 1:

Census 119	Positive and negative correlation 123
Scatterplot 121	Simple random sample 124
Regression line 123	Stratified random sample 125
Correlation coefficient 123	

Key Research Studies

You should be familiar with both the methodology and the results of these research studies.

> Darley and Latenè: bystander apathy 97
> Hirschi and Stark: delinquency and religious commitment 99
> Stark and others: church membership, geographical area, and delinquency 103
> Paige: comparative study of family systems of primitive societies 113

Key Theories

You should be able to explain the assumptions of the theories and, when applicable, cite related research findings.

> Macro sociology theories (in general)
> Functionalism
> Social evolution (Lenski)
> Conflict theory (Marx)

Completion

1. The results of the research by Darley and Latenè found that the larger the group believed to be present, the _____ an individual will feel personal responsibility to act in an emergency.

2. In _____, the data are collected by personal interview or questionnaire.

3. A relationship is _____ if it disappears when some third variable is controlled.

4. Macro sociological theories attempt to explain the existence of _____.

5. _____ sociologists assume that societies are systems.

6. Because the parts of a system are interdependent, they tend to fall into some kind of _____ or balance.

7. The five basic institutions in any society include religion, the political order, _____, _____, and _____.

8. _____ are clusters of specialized roles, groups, organizations, customs, and activities devoted to meeting social _____.

9. One adult couple and their children constitute a(n) _____ family.

10. Another structure by which the same function can be accomplished is termed a(n) _____.

11. Arrangements among structures that harm or distort the system are termed _____.

12. _____ theories suggest that societies with structures that enable them to adapt to their physical and social environments have a better chance for survival than do societies that fail to develop such structures.

13. A faulty assumption made by nineteenth-century social evolutionists was that social change is _____ and progressive.

14. _____ theorists ask how social structure serves the interests of various competing groups within a society.

15. An ethnic group is a good example of a(n) _____ group.

16. In cultures that possess a(n) _____ rule of residence, newlyweds reside near or with the bride's family.

17. Paige's research demonstrated a very strong correlation between rules of residence and _____.

*18. If one variable rises while the other falls, there is a(n) _____ correlation.

*19. When correlations are _____, the regression line slopes from lower left to upper right.

*20. Sampling that proceeds through a series of levels is termed _____ sampling.

*Drawn from Special Topic 1.

Multiple-Choice

1.   Research by Darley and Latenè found that:
        a.   the larger the group present, the more an individual will feel personal responsibility to act in an emergency.
        b.   the larger the group present, the less an individual will feel personal responsibility to act in an emergency.
        c.   the size of the group present has no effect on bystander apathy.
        d.   church attendance and delinquency rates are inversely correlated.
        e.   none of the above

2.   Survey research:
      a.   collects data using questionnaires and personal interviews.
      b.   uses samples.
      c.   can establish time order more easily than can experiments.
      d.   a and b
      e.   all of the above

3.   When sex differences were controlled, Hirschi and Stark found that:
      a.   boys who attend church are less likely to be delinquents than boys who
            did not attend church.
      b.   girls who attend church are less likely to be delinquents than girls who
            did not attend church.
      c.   the more people present at an emergency, the more likely the victim is to
            receive help.
      d.   a and b
      e.   none of the above

4.   Research on delinquency and religion has found that:
      a.   religion is negatively correlated with delinquency in schools where the
            majority of the students are religious.
      b.   religion is positively correlated with delinquency in schools where the
            majority of the students are religious.
      c.   religion is negatively correlated with delinquency in schools where most
            students are not religious.
      d.   religion is positively correlated with delinquency in schools where most
            students are not religious.
      e.   there was no correlation between religion and delinquency in schools
            where most of the students are religious.

5.   Elements of systems include:
      a.   separate parts or structures.
      b.   interdependence among the parts.
      c.   equilibrium among the parts.
      d.   b and c
      e.   all of the above

6.   Basic institutions found in all societies include:
      a.   the family.
      b.   the political order.
      c.   religion.
      d.   a and c
      e.   all of the above

7.   Macro sociologists assume that:
        a.   societies are never wholly static.
        b.   every structure is related to every other structure.
        c.   the same degree of interdependence among structures exists in all societies.
        d.   b and c
        e.   all of the above

8.   _____ theories explain social structures on the basis of their consequences for other parts of the system.
        a.   Conflict
        b.   Functional
        c.   Social evolution
        d.   Symbolic interaction
        e.   Choice

9.   Arrangements among structures that harm or distort the system are termed:
        a.   functional alternatives.
        b.   functional requisites.
        c.   dysfunctions.
        d.   latent functions.
        e.   manifest functions.

10.   Contemporary evolutionary theories:
        a.   always assume that all societies evolve to more complex cultures.
        b.   are meant to apply to the population of cases.
        c.   assume that all change is inevitable and progressive.
        d.   all of the above
        e.   none of the above

11.   _____ theorists ask how social structure serves the interests of various competing groups within a society.
        a.   Functional
        b.   Conflict
        c.   Social evolutionary
        d.   Micro
        e.   Choice

12.   The term *status group* is most closely associated with:
        a.   Marx.
        b.   Weber.
        c.   Lenski.
        d.   Paige.
        e.   Darley.

13. Macro sociological research must always be based on:
   a. individuals.
   b. the comparative study of groups.
   c. case studies.
   d. primitive societies.
   e. none of the above

14. Societies with factional politics:
   a. reach decisions through competition and conflict.
   b. stress agreement rather than disagreement.
   c. contain internal groups that stress their own interests.
   d. a and c
   e. none of the above

15. When Paige examined primitive societies, he found that:
   a. kinship and residence are primary bases for group formation.
   b. conflict occurs primarily among men of different kinship groups.
   c. patrilocal societies were often more communal than matrilocal societies.
   d. a and b
   e. all of the above

16. Societies in which the bride leaves home after marriage and the couple takes up residence with or close to the husband's family have a _____ rule of residence.
   a. matrilocal
   b. neolocal
   c. patrilocal
   d. fratralocal
   e. nuclear

17. Paige's study of structure and conflict used:
   a. field research.
   b. the experimental method.
   c. comparative research.
   d. simple random samples.
   e. none of the above

*18.   Which of the following is/are true?
a   Correlations can be positive or negative.
b.   When there is no correlation, the regression line slopes from upper left to lower right.
c.   The closer the correlation coefficient is to 1.00, the less the correlation between the two measures.
d.   a and c
e.   all of the above

*19.   The odds that a sample will be like the whole population depend on:
a   the absolute size of the sample.
b.   the ratio of the sample size to the population size.
c.   the use of stratified rather than simple random samples.
d.   a and b
e.   all of the above

20.   Which of the following is not correctly paired with his or her contribution?
a   Darley and Latenè: bystander apathy
b.   Karl Marx: social evolutionary theory
c.   Max Weber: status group
d.   Paige: comparative study of family systems of primitive societies
e.   b and c

*Drawn from Special Topic 1.

Essay

1.   A.   Explain functional and conflict theories. (comprehension)
       B.   Attempt to integrate functional and conflict theories into a more general explanation of society. (application)
       C.   *Compare and contrast* functional and conflict theories. (analysis)

2.   A.   Define micro and macro sociology. (knowledge)
       B.   Explain the relationship between micro and macro sociology. (comprehension)
       C.   Show how the research on religion and delinquency illustrates this relationship. (application)

3.   Briefly discuss the research findings of Stark and others on religion and delinquency.

4.   Using examples, discuss the three elements of a system.

5. Discuss Paige's research on kinship structure and political conflict.

6. Newspapers frequently report research findings taken out of context with little background information about the specifics of the study. This often gives the uninformed public a false image of causality where it may not exist.

    Assume you read an article that reported that college students who attend church regularly are less likely to engage in pre-marital sex than those who don't. You become interested in this finding and read the actual study in a research journal. There you learn the study compared students in a small private religious college where most students lived at home with those in a large residential public college. Based on your knowledge of the relationship between delinquency and church attendance and the importance of social structure discussed in this chapter, what can you conclude about the relationship between church attendance and pre-marital sex that was omitted in the newspaper?

Answers

Completion

1.	less	11.	dysfunctions
2.	survey research	12.	Social evolutionary
3.	spurious	13.	inevitable
4.	social structures	14.	Conflict
5.	Macro	15.	status
6.	equilibrium	16.	matrilocal
7.	education, family, the economy	17.	political conflict
8.	Institutions, needs	18.	negative
9.	nuclear	19.	positive
10.	functional alternative	20.	stratified random

Multiple-Choice

1.	b	11.	b
2.	d	12.	b
3.	e	13.	b
4.	e	14.	d
5.	e	15.	d
6.	e	16.	c
7.	a	17.	c
8.	b	18.	a
9.	c	19.	a
10.	b	20.	b

CHAPTERS 1 TO 4

REVIEW AND SPECIAL PROJECT

Review

Chapters 1 through 4 have introduced sociology, discussing it as a science and introducing many new concepts central to sociology. They have discussed both micro and macro sociology and the theory and research in these areas.

You may wish to test your knowledge of this material by actually trying your hand at conducting some sociological research. Although it might be difficult to conduct a study in a few weeks without administrative and financial backing, it is possible to design one. Through this design, you will test your knowledge of several areas and obtain firsthand knowledge of sociology in action.

Special Project

Using students on your campus as subjects, assume that you want to replicate one of the following studies: Ofshe's study of attachments and conformity, Milgram's "small world" research, or one of the studies on religion and delinquency conducted by Stark and others. Design a study based on the original study you have chosen.

As you design your study, be certain to address the following questions:

1.   What did the original researcher hypothesize and what were his or her results?
2.   What do you hypothesize? Do you expect similar results?
3.   Is your research experimental or nonexperimental?
4.   How will you draw your sample? What kind will you use?
5.   What method will you use to gather your data?
6.   How will you establish the criteria of causation?
7.   Is your study more in the realm of micro sociology or macro sociology?
8.   Does your study use the assumptions of any particular theory? What are they?
9.   Can you identify any problems or issues that might arise if you actually were to conduct this study? How might you address them in advance?

CHAPTER 5

BIOLOGY, CULTURE, AND SOCIETY

Overview

Chapter 5 starts with a brief historical discussion of instinctual and environmental theories of behavior. It explains behavioral genetics and shows how most social scientists today take a balanced position between heredity and environment. Research, such as numerous twin studies, is cited to support the notion of interplay between biology and environment. The chapter then turns its attention to recent findings on the relationship between hormones and behavior and the Vietnam Veteran Study is highlighted in "a closer view." The chapter then focuses on a discussion of research findings from ethology on learned behavior among animals. Jane Goodall's pioneering research with chimpanzees is described, and the chapter closes with a discussion of symbolic communication in primates.

Capsule Summary

Early in this century, <u>instinctual theories</u> dominated the social sciences. They postulated that behavior was <u>inborn</u> and the result of <u>heredity</u>. By the 1930s, however, <u>environmental theories</u> dominated. They postulated that behavior was strictly the result of <u>cultural</u> and <u>social</u> influences, and that <u>biology</u> played no role.

Today most social scientists assume that humans are the result of the interplay between their <u>biology</u> and their <u>social</u> and <u>cultural</u> environment. (The concepts <u>genotype</u> and <u>phenotype</u> illustrate this interplay.) Indeed, the rapidly growing field of <u>behavioral genetics</u> seeks to identify <u>traits</u> that <u>influence behavior</u> and have some <u>genetic basis</u>. Research in this area has emphasized the study of <u>identical twins,</u> particularly identical twins <u>reared apart</u>. These twins often exhibit <u>similar although not identical character-istics</u>. Similarly, the interplay of heredity and environment is illustrated by the research on the <u>increasing physical size</u> of Americans since the turn of the century. <u>Environmental conditions</u> can suppress or enhance <u>genetic potential</u>.

Recent research has focused on the relationship between <u>hormones</u> and behavior. Studies have shown a relationship between <u>testosterone</u> <u>levels</u> and <u>impulsiveness, violent crime,</u> and <u>adolescent sexual activity</u> although these studies were based on limited samples. The <u>Vietnam Veteran Study</u> found that high levels of <u>testosterone</u> in <u>men</u> are related to problems such as <u>alcoholism, drug use, trouble with the law,</u> and <u>wife beating</u>.

Sociologists have long assumed that humans differ from other animals in that we <u>alone possess culture and language</u>. Recent studies in <u>ethology</u> by field researchers such as <u>Jane Goodall</u> have discovered evidence of <u>learned behavior, toolmaking,</u> and use of <u>symbolic communication</u> among other animals, particularly primates. The <u>Harlows'</u> research with <u>monkeys</u> showed that <u>early isolation</u> had <u>detrimental effects</u> on adult <u>behavior</u> such as <u>sexual performance</u> and <u>social relationships</u>. A landmark in the <u>study of communication</u> was reached when the <u>Gardners</u> taught a chimp, <u>Washoe</u>, American

Sign Language. Washoe was able to communicate through sign language and eventually to teach these signs to another infant chimp placed in her care.

Key Concepts

You should be prepared to explain the concepts and terms listed here as well as give several examples of each.

Instinct 129	Phenotype 133
Gene 132	Hormone 138
Chromosome 132	Testosterone 138
Genotype 132	Estrogen 138
	Tool 143

Key Research Studies

You should be familiar with both the methodology and the results of these research studies.

Rosenthal, Schuckit, and others: twin studies 134

Daitzman and Zuckerman, Dobbs and others: relationship between testosterone levels and anti-social behavior 139

Vietnam Veteran Study: relationship between testosterone levels and anti-social behavior 139

Goodall: field research with chimpanzees 140

Harlow and Harlow: effects of isolation on infant monkeys 144

Gardner and Gardner: teaching American Sign Language to chimps 145

Key Theories

You should be able to explain the assumptions of these theories and, when applicable, cite related research findings.

Instinctual theory (McDougall)

Environmental theory

Behavioral genetics

Completion

1. A(n) _____ is a form of behavior that occurs in all normal members of a species without having been learned.

2. The major proponent of instinctual theories was social psychologist _____.

3. By the 1930s, the social sciences were dominated by purely _____ theories.

4. Today we take the position that human beings are the result of the interplay between their _____ and their _____.

5. In our chromosomes are tiny structures termed _____ that contain DNA.

6. The phenotype is the actual outcome of the interplay between the _____ and the environment.

7. Research has found that Americans are growing both _____ and _____ than did earlier generations.

8. The male sex hormone is termed _____.

9. Research on men in prison found _____ levels of testosterone in men convicted of violent crimes than those convicted of non-violent crimes.

10. The Vietnam Veteran Study found the higher their level of testosterone, the _____ likely men are to physically abuse their wives.

11. The Vietnam Veteran Study found the _____ their level of testosterone, the less likely men are to become highly educated.

12. A(n) _____ is an object that has been modified to suit a particular purpose.

13. Older introductory texts have often argued that humans differ from other animals in that only humans possess _____ and _____.

14. Students of animal behavior are termed _____.

15. Research on animal behavior has found that while much behavior is instinctual, there is increasing evidence that some behavior is _____.

16. The Harlows' study of monkeys found that many of the effects of early isolation seem _____.

17. Goodall discovered both toolmaking and _____ among the chimps in the wild.

18. The Gardners taught Washoe to communicate through the use of _____.

19. It is widely recognized that much nonhuman behavior is _____, although much is also learned.

20. A virtue of animal studies is that we can manipulate _____ to study adaptation.

Multiple-Choice

1. Early instinctual theories:
    a. discounted the impact of cultural and social influences on human development.
    b. dominated social science during the 1930s.
    c. assigned no role to heredity.
    d. b and c
    e. all of the above

2. The major proponent of instinctual theories was:
    a. Harry Harlow.
    b. Alfred Binet.
    c. William McDougall.
    d. James Tanner.
    e. Jane Goodall.

3. Today most social scientists take the premise that human development results from:
    a. purely environmental influences.
    b. purely biological influences.
    c. an interplay between biology and environment.
    d. purely cultural influences.
    e. none of the above

4. The sum total of the genetic instructions that an organism receives from its parents is called the:
    a. genotype.
    b. chromosomal number.
    c. phenotype.
    d. genonumber.
    e. phenonumber.

5. Behavioral geneticists have claimed considerable success in isolating human characteristics and behavior that are influenced to a substantial degree by genetic inheritance, including:
    a. alcoholism.
    b. intelligence.
    c. a tendency toward impulsive and aggressive behavior.
    d. a and c
    e. all of the above

6. Studies of identical twins reared apart have found their IQs to be:
    a. identical.
    b. extremely similar.
    c. markedly different.
    d. unmeasurable.
    e. none of the above; no such subjects have been discovered.

7. Studies based on identical twins have produced _____ of a genetic factor in a wide range of personality traits.
    a. no evidence
    b. weak evidence
    c. strong evidence
    d. untestable evidence

8. In the past, environmental suppressors of humans' natural growth patterns included:
    a. inadequate nutrition.
    b. a diet containing too much meat and dairy products.
    c. chronic poor health.
    d. a and c
    e. all of the above

9. Changes brought about by industrialization included:
    a. declines in infant mortality and doubled life expectancy.
    b. increased size of people.
    c. earlier maturation of people.
    d. all of the above
    e. none of the above

10. Until very recently, connections between various hormones and behavior were little studied because:
   a. it has been difficult and expensive to measure variations in hormonal levels.
   b. those qualified to study hormones were not trained to study behavior.
   c. those trained to study behavior were not trained to study hormones.
   d. a and b above
   e. all of the above

11. Research by Daitzman and Zuckerman found that:
   a. men with high levels of testosterone scored higher on tests of impulsiveness and sensation seeking.
   b. men with low levels of testosterone scored higher on tests of impulsiveness and sensation seeking.
   c. there was no relationship between testosterone levels and conviction rates for violent crimes for men.
   d. men with high levels of testosterone were more likely to attain high status positions than those with lower levels.
   e. none of the above

12. Research findings by Dabbs, Julian and McKenry, and Udry were limited in applicability because:
   a. only men were used in the studies.
   b. the researchers were not trained to study hormones.
   c. both males and females were studied and females do not have testosterone.
   d. the sample size was relatively small.
   e. none of the above

13. The Vietnam Veteran Study found that the higher their level of testosterone, the _more_ likely men are to:
   a. engage in extramarital sex.
   b. become highly educated.
   c. have problems with alcoholism.
   d. a and c above
   e. all of the above

14. The Vietnam Veteran Study found that men with high levels of testosterone are more likely to:
   a. get in trouble with the law as an adult.
   b. have been punished while in military service.
   c. be unemployed.
   d. all of the above
   e. none of the above

15. The Vietnam Veteran Study found that the higher their level of testosterone, the less likely men are to:
    a. get divorced.
    b. become highly educated.
    c. obtain a high status occupation.
    d. b and c above
    e. a and c above

16. Ethologists study:
    a. primitive societies.
    b. the interplay between environment and biology in humans.
    c. animal behavior.
    d. genetic potential in humans.
    e. none of the above

17. The Harlows found that monkeys reared in total isolation:
    a. were no different from monkeys reared with other monkeys.
    b. made up for early deficits when they joined monkey colonies.
    c. were unable to engage in social and sexual relationships.
    d. a and b
    e. none of the above

18. Jane Goodall discovered that chimpanzees in the wild:
    a. make and use tools.
    b. communicate through sign language.
    c. kill other chimps.
    d. a and c
    e. all of the above

19. The advantages of primate studies are that:
    a. we can manipulate the environment to study adaptation.
    b. we can observe primates as substitutes for our most ancient ancestors.
    c. we can seek basic elements of social organization by examining societies that have not been overlaid with a great deal of culture.
    d. all of the above
    e. none of the above

20. The case of Washoe illustrates that:
    a. chimp behavior is determined entirely by instincts.
    b. chimps can learn sign language but cannot use it to form sentences.
    c. chimps cannot transmit language to their young.
    d. chimps can learn to speak words.
    e. none of the above

Essay

1.   A.   Distinguish between genotypes and phenotypes. (knowledge)
     B.   Explain the following statement: "Human beings are a result of the interplay between their biology and their social and cultural environment." (comprehension)
     C.   Give an example of a human behavior or characteristic and show how it is the result of the interrelationship between genetic inheritance and the environment. (application)

2.   A.   Explain instinctual and environmental theories. (comprehension)
     B.   Using current research, interrelate the role of heredity and environment. (application)
     C.   Contrast instinctual and environmental theories. (analysis)

3.   Discuss some of the research findings about the relationship between testosterone levels and behavior.

4.   Describe some of the societal reasons why we are growing bigger and faster than did our ancestors.

5.   Using the research studies on primates, show how these research findings do not support the premise that humans alone possess culture and can communicate through a formal system.

6.   Teenage pregnancy, especially among young teens, is considered a serious social problem in today's society. How can the current high rates of early teen pregnancy be in part explained by the growth revolution? In addition, how have cultural changes contributed to this problem? Can you think of other areas in which the growth revolution has contributed to changes in the life styles of young adolescents? Again, what cultural factors may also be involved?

Answers

Completion

1. instinct
2. William McDougall
3. environmental
4. biology, social and cultural environment
5. genes
6. genotype
7. faster, larger
8. testosterone
9. higher
10. more
11. higher
12. tool
13. culture and language
14. ethologists
15. learned
16. irreversible
17. meat eating (killing)
18. American Sign Language
19. instinctive
20. environment

Multiple-Choice

1. a
2. c
3. c
4. a
5. e
6. b
7. b
8. d
9. d
10. e
11. a
12. d
13. d
14. d
15. d
16. c
17. c
18. e
19. d
20. c

CHAPTER 6

SOCIALIZATION AND SOCIAL ROLES

Overview

Chapter 6 introduces the process of socialization and describes research findings on the effects of early isolation and deprivation as evidence of the importance of this process. Piaget's work on cognitive development is discussed in-depth. Brown and Bellugi's work on language acquisition is also included in the discussion of cognitive structures. The relationship between interaction and cognitive development with an emphasis on the attachment-teaching hypothesis is discussed, as is research on the relationship between cognitive development and children of single mothers and children in day care. Emotional development, the emergence of the self, and personality development are described. The chapter then turns its attention to the theory of cultural determinism and its assumptions, research findings, and recent criticisms. Margaret Mead's famous works provide examples of extreme cultural determinism. Differential socialization is discussed, and studies by Kohn and others are described in-depth. Goffman's "stage analogy" of interaction is described with examples of its concepts. The chapter closes with a discussion of the origins and current status of differential socialization on the basis of gender.

Capsule Summary

Socialization is the crucial learning process that allows us to possess culture and participate in social relations. Feral children are examples of extreme isolation and deprivation. Early studies such as those by Skeels and Dye documented the importance of interaction and contact during the early years.

For many years psychology was dominated by the stimulus-response (SR) theory of learning, which argued that behavior was a response to external stimuli and that learning was a result of reinforcement. Piaget took issue with the SR theory and argued that the human mind develops and functions on the basis of cognitive structures. Through extensive research, he proposed the existence of four stages of cognitive development: sensorimotor, preoperational, concrete operational, and formal operational. Children in these stages differ in their ability to comprehend concepts and situations. Others who endorsed the concept of cognitive structures include Brown and Bellugi, whose research on language acquisition indicated that young children's speech often seems to indicate a search for grammatical rules.

Piaget has been criticized for ignoring the role of social interaction in his work. Recent research has focused on the relationship between verbal interaction and language acquisition (the attachment-teaching hypothesis), and this research has yielded surprising results. Research findings on the relationship between cognitive development and children of single mothers and children in day care have yielded mixed results.

An important aspect of the socialization process is the emergence of the self. Research by Bain, Flavell, and others has expanded on earlier work by Piaget and Mead. Personality refers to the consistent pattern of thoughts, feelings, and actions displayed by an individual. Personality emphasizes both the similarities and differences between and among individuals.

Cultural determinism was a dominant theory in anthropology in the 1920s and 1930s. In its extreme form, cultural determinism argues that personality is totally shaped by culture and that child-rearing practices are critical in determining later personality characteristics. Mead's works on adolescence in Samoa and sex and temperament are classic examples of cultural determinism. These studies have recently come under criticism, and today most social scientists take the position that culture and early socialization are important in shaping personality but are not the only factors.

Because not all people in a society are expected to play identical roles, not all members of a society are socialized exactly the same. Differential socialization accounts for some of the differences among people. Work by Kohn et al., for example, found that parents often socialize their children on the basis of the roles that they expect them to perform. These expectations often reflect the parents' own working conditions.

Erving Goffman studied interaction from the point of view that the world is a stage upon which we are all actors. He distinguished between role and role performance and used concepts such as props, front stage and backstage, and impression management in his analogy.

Differential socialization is possibly best illustrated in sex-role socialization. Although the sexes differ biologically, these differences were a more important basis for the assignment of roles in primitive societies than they are at present. These early societies laid the foundation for differential socialization by gender that has become a part of our culture. Today we still socialize males and females differently. Research by DeLoache et al. found that when reading to their children, mothers assign gender to neutral characters in children's books. These assignments are typically based on conventional sex-role stereotypes. Likewise, Richner found that gender plays a significant role in older children's choice of playmates.

Key Concepts

You should be able to explain the concepts listed here and be able to cite several examples of each.

Feral children 151	Differential socialization 165
Socialization 152	Role 166
Cognitive structure 155	Adult socialization 168
Sensorimotor stage 156	Longitudinal study 168
Rule of object permanence 156	Role performance 170
Preoperational stage 156	Impression management 170
Concrete operational stage 156	Stage 170
Rule of conservation 156	Backstage 170
Formal operational stage 156	Teamwork 170
"Motherese" 158	Studied nonobservance 170
Self 159	Deviant role 170
Personality 161	Sex-role socialization 171
Cultural determinism 162	

Key Research Studies

You should be familiar with both the methodology and the results of these research studies.

Skeels and Dye: early study of the effects of socialization on retardation 153

Brown and Bellugi and others: language acquisition of children 157

Bretherton et al.: "attachment-teaching hypothesis" 158

Desai, Belsky, and others: Cognitive development, single mothers, and day care 159

Bain and others: the emergence of the self 160

Margaret Mead: anthropological studies supporting cultural determinism—*Coming of Age in Samoa: Sex and Temperament* 162

Melvin Kohn et al.: relationships among social class, parental expectation of role, and socialization 167

Goffman: performing social roles 169

Data on gender roles and gender socialization in premodern societies 172

DeLoache et al.: assignment of gender to neutral characters in children's books 175

Richer: games and gender 176

Key Theories

You should be able to explain the assumptions of these theories and, when applicable, to cite related research findings.

Stimulus-response

Theory of cognitive stages (Piaget)

Cultural determinism (Boaz and Mead)

Completion

1. Children who are neglected and isolated from human contact are termed _____.

2. The learning process by which infants become normal human beings, possessed of culture and able to participate in social relations, is termed _____.

3. Socialization related to roles and to _____ is termed *differential socialization.*

4. The theory that behavior is merely a response to external stimuli and that we repeat whatever behavior has been reinforced by our environment is termed _____.

5. Piaget argued that the human mind develops and functions on the basis of _____ or general rules for reasoning.

6. Piaget's cognitive stages include the _____, preoperational, concrete operational, and _____.

7. The rule of _____ is Piaget's term for the principle that objects continue to exist even when they are out of sight.

8. The third stage in Piaget's theory of cognitive development is the _____.

9. Researchers have concluded that perhaps _____ of all adults do not reach the formal operations stage.

10. Brown and Bellugi found that young children experiment with speech in ways that appear to involve a search for _____.

11. An individual's consistent pattern of thoughts, feelings, and actions is termed his or her _____.

12. _____ published and taught the theory of cultural determinism.

13. Mead described the _____ as gentle, unaggressive, and passive and argued that both men and women have "feminine temperaments."

14. Societies faced with chronic warfare will place exceptional value on _____ infants.

15. Kohn found that in contrast to working-class parents, middle-class parents were more concerned about their children being capable of _____ and _____.

16. A study in which observations are made of the same people at several different times is termed a(n) _____ study.

17. The actual conduct of a particular individual while on duty in a position is termed
_____.

18. Goffman termed "the conscious manipulation of role performance" _____.

19. A set of norms attached to a position that, in turn, violates the norms adhered to by the larger society is termed a(n) _____.

20. DeLoache et al. found that mothers were most likely to attribute _____ identities to gender-neutral characters.

Multiple-Choice

1.   The socialization process:
   a.   is a learning process.
   b.   begins at birth and ends at age 7.
   c.   literally means to be "made social."
   d.   a and c
   e.   all of the above

2.   Skeels and Dye found that infants who had been placed under the personal care of an older girl showed _____ when compared with those who remained in the orphanage.
   a.   slight improvement
   b.   dramatic improvement
   c.   no improvement
   d.   slight decline in ability
   e.   marked decline in ability

3.   The stimulus-response theory postulated that learning results from:
   a.   reinforcement.
   b.   cognitive structures.
   c.   developmental stages.
   d.   instincts.
   e.   heredity.

4.   The first stage of cognitive development is the:
   a.   preoperational.
   b.   formal operational.
   c.   sensorimotor.
   d.   concrete operational.
   e.   sensorioperational.

5.   During the preoperational stage, children lack:
      a.   the rule of object permanence.
      b.   the rule of conservation.
      c.   the ability to put themselves in someone else's place.
      d.   b and c
      e.   all of the above

6.   Research on the relationship among cognitive development, single mothers, and day care has:
      a.   consistently found that day care seriously hampered mother-child interaction and cognitive development.
      b.   found that the mother's own cognitive development score is a much better predictor of her child's score than her age and marital status.
      c.   consistently found no relationship between day care and cognitive development.
      d.   found that a mother's age and marital status were a better predictor of her child's score than her own cognitive development score.
      e.   none of the above

7.   In their study of language acquisition, Brown and Bellugi discovered that:
      a.   young children's speech contains only the most vital words.
      b.   parents frequently echo their children, thus expanding and correcting their sentences.
      c.   young children experiment with speech in ways that seem to involve a search for grammatical rules.
      d.   a and c
      e.   all of the above

8.   Research in the area of the attachment-teaching hypothesis has shown that:
      a.   children who are exposed to "motherese" acquire language much more rapidly than those who are not.
      b.   children who are talked to learn to talk sooner and better.
      c.   the degree of attachments a child enjoys plays a major role in language acquisition.
      d.   all of the above
      e.   none of the above

9.   In terms of their personalities:
      a.   all humans are alike in some ways.
      b.   all humans are like only some other humans.
      c.   all humans are unique in some ways.
      d.   b and c
      e.   all of the above

10. The principle of cultural determinism argues that individuals' personalities:
    a. are tiny replicas of their cultures.
    b. are the result of an interplay between biology and culture.
    c. are entirely the result of heredity.
    d. are not well developed among primitive people.
    e. none of the above

11. The theory of cultural determinism is most closely associated with:
    a. Jean Piaget.
    b. Franz Boaz.
    c. Roger Brown.
    d. Melvin Kohn.
    e. Erving Goffman.

12. Margaret Mead argued that the differences in temperament between the Arapesh and the Mundugumor were the result of:
    a. innate biological differences.
    b. child-rearing practices, especially during infancy.
    c. the physical location of their respective societies.
    d. b and c
    e. none of the above

13. In his *initial* study, Kohn found that middle-class parents typically stressed the value(s) of:
    a. self-expression.
    b. independence.
    c. conformity.
    d. a and b
    e. all of the above

14. In later studies, Kohn et al. have found that in *both* the United States and Poland:
    a. the parental child-rearing values of parents depend on their own occupational experiences.
    b. parental values were very effectively transmitted by parents to their children.
    c. in contrast to U.S. fathers, Polish fathers have much more influence on their children than do mothers.
    d. a and b
    e. all of the above

15.   In a longitudinal study, observations are made of:
        a.   different people at several different times.
        b.   different people at the same time.
        c.   the same people at several different times.
        d.   nonhuman animals in natural settings.
        e.   none of the above

16.   Goffman termed the "conscious manipulation of scenery, props, costumes, and
        our behavior in order to convey a particular role image to others":
        a.   role failure.
        b.   studied nonobservance.
        c.   impression management.
        d.   backstage behavior.
        e.   teamwork.

17.   Pretending not to see miscues in others' role performance is called:
        a.   role failure.
        b.   studied nonobservance.
        c.   impression management.
        d.   backstage behavior.
        e.   teamwork.

18.   Juhasz's study of self-esteem of seventh- and eighth-graders found that _____
        has a potent effect on self-conceptions:
        a.   gender
        b.   academic ability
        c.   parental occupation
        d.   religion
        e.   none of the above

19.   Research by DeLoache et al. found that the subjects in the study:
        a.   typically referred to gender-neutral characters as males.
        b.   typically referred to gender-neutral characters as females.
        c.   always identified the "teachers" as female and the "drivers" as male.
        d.   a and c
        e.   none of the above

20.   Richer's results suggest that:
       a.   older children display gender preferences in their play but that younger children do not.
       b.   older children do not display gender differences in their play.
       c.   the gender preferences of slightly older children seem to be innate.
       d.   a and c
       e.   none of the above

Essay

1.   A.   List Piaget's four stages of cognitive development. (knowledge)
      B.   Describe these stages using examples. (comprehension)
      C.   Contrast the type of thinking characteristic of a child in the preoperational stage with that of a child in the concrete operations stage. (analysis)

2.   A.   Explain the stimulus-response theory of learning. (knowledge)
      B.   Show how the work of Piaget or of Brown and Bellugi does not support the assumptions of S-R theory.
      C.   Design a study to test Piaget's theory or to replicate the work of Brown and Bellugi. (application)

3.   Explain cultural determinism. Discuss Mead's studies as an example of extreme cultural determinism.

4.   Discuss some of the recent criticisms of Mead's work. Explain the contemporary position held by most sociologists regarding the relationship between culture and personality.

5.   Explain differential socialization. Discuss the work of Kohn et al. in this area.

6.   If you have already chosen a college major and potential career, what factors in your socialization experience may have contributed to your choice? If you are as yet somewhat undecided about a major or career, are there any areas you are seriously considering or any areas you have ruled out? Why? (Remember socialization may be a very subtle process.) Since socialization is a lifelong process, can you think of any experiences or influences that may cause you to change or reconsider your plans?

Answers

Completion

1. feral children
2. socialization
3. role expectations
4. stimulus-response
5. cognitive structures
6. sensorimotor, formal operational
7. object permanence
8. concrete operational
9. one-half (50%)
10. grammatical rules
11. personality
12. Boaz
13. Arapesh
14. male
15. self-expression, independence
16. longitudinal
17. role performance
18. impression management
19. deviant role
20. male (or masculine)

Multiple-Choice

1. d
2. b
3. a
4. c
5. d
6. b
7. e
8. b
9. e
10. a
11. b
12. b
13. d
14. d
15. c
16. c
17. b
18. a
19. a
20. a

CHAPTER 7

CRIME AND DEVIANCE

Overview

Chapter 7 opens with a brief description of deviance and focuses specifically on the type of deviance known as crime. It discusses some of the limitations of traditional definitions of crime and offers a definition relatively free of legal and political specifications. It discusses the characteristics of ordinary crime and how these differ from most crimes reported in the media. It then turns its attention to early and contemporary biological and psychological theories and related research findings. Sociological theories are introduced with a brief discussion of the importance of attachments. The chapter then discusses in-depth the major sociological theories of deviance: differential association/social learning, subcultural deviation, structural strain, control theory, anomie, and labeling. In each case, the major assumptions of the theory are considered, relevant research findings are cited, and theoretical shortcomings and criticisms are noted. The chapter concludes with a discussion of the relationship between drug use and crime.

Capsule Summary

Behavior that violates norms is termed <u>deviant behavior</u>. <u>Crime</u>, the most serious type of deviance, is defined as <u>"acts of force or fraud undertaken in pursuit of self interest."</u> Recent research on <u>robbery</u>, <u>burglary</u>, and <u>homicide</u> has found most crimes are <u>easy to commit</u> and very <u>simple in design</u>. They typically involve <u>little planning</u>, are <u>brief in duration</u>, and <u>rewards</u>, while <u>small</u>, are <u>immediate</u>.

       There are sociological and nonsociological theories of deviation. <u>Biological</u> theories attempt to show how deviants differ physically from nondeviants. <u>Lombroso's theory</u> of "born criminals" and the more recent approach of <u>Gove (age, gender, biology, and deviance)</u> are examples of biological explanations. Similarly, most <u>psychological research</u> has been unable to find a significant relationship between <u>personality</u> type and <u>deviance</u>, although recent studies do indicate a link between <u>criminal actions</u> and <u>low self-control</u>.

       <u>Sociological</u> theories of deviance include <u>differential association/social learning</u>, <u>subcultural deviance</u>, <u>structural strain</u>, <u>control theory</u>, <u>anomie</u>, and <u>labeling</u>. Each theory is based on different assumptions, and each has certain shortcomings.

       <u>Sutherland's theory of differential association</u> argues that deviant behavior, like other behavior, is learned through socialization. Thus, attachments to others who are deviant may encourage deviation. Later learning theories have included the concept of <u>selective reinforcement</u> to further explain this process.

       <u>Subcultural deviance</u> emphasizes the <u>conflicts over norms</u>, which may arise in a society that contains many subcultures. Thus, behavior that may be conforming in a deviant subculture may be deviant to the general culture.

       <u>Merton's theory of structural strain</u> argues that <u>deviance results</u> from the <u>frustrations</u> experienced by those who occupy <u>disadvantaged positions</u> in the

stratification system. It argues that in attempting to conform to culturally approved goals, the poor will find the legitimate means to these goals blocked and, hence, will resort to illegitimate deviant means to attain them. A major shortcoming of strain theory is that it is unable to explain high rates of white-collar crime in society. Indeed the very definition of white-collar crime has caused difficulties for social scientists. Shapiro has argued that a new concept of crimes involving violations of trust should be substituted for this term.

Attempting to explain conformity rather than deviance, control theory argues that conformity is tied to the bonds between an individual and the group. If the bonds are strong, then the individual is more likely to conform. It recognizes the existence of four types of bonds—attachments, involvements, investments, and beliefs—and argues that an individual is likely to deviate when these bonds are weak.

Anomie (literally, "normlessness") argues that deviance is the result of low social and moral integration. This idea was first proposed by Durkheim in his discussion of "moral communities." Recent research has supported Durkheim's notions about the relationship between crime and moral and social integration.

Labeling theory distinguishes between primary and secondary deviation and focuses on the effect that a deviant label has on both the recipient and on society. It argues that labels are not uniformly applied to all deviants and that ultimately the label itself may cause a return to deviant behavior.

Recent research has also shown a relationship between weather and crime rates. Crime rates tend to be highest in the warmest summer months which provide the greatest opportunities to commit crimes. There is also evidence of a strong relationship between drug use and crime, although most drug use does not cause crime but rather is an aspect of the offender lifestyle.

Key Concepts

You should be able to explain the concepts listed here and be prepared to cite several examples of each.

Deviance 181	Investment 205
Crime 182	Involvement 205
Offender versatility 186	Belief 205
Born criminals 188	Internalization of norms 205
Subcultural deviance 196	Anomie 208
White-collar crime 200	Moral community 208
Crimes involving violations	Social integration 209
of trust 201	Moral integration 209
Conformity 202	Primary deviance 211
Attachment 202	Secondary deviance 211

Key Research Studies

You should be familiar with both the methodology and the results of these research studies.

Gottfredson and Hirschi: characteristics of "ordinary crime" 182
Lombroso: early study of "born criminals" 188
Gove: age, gender, biology, and deviance 189
Dabbs and others: testosterone levels and violent crime 192
Link, McFarland and others: mental illness and crime 193
Gottfredson and Hirschi: criminal acts and self-control 193
Linden and Fillmore: delinquency study combining differential association and
control theory 206
Lebeau and others: weather and crime rates 210
Gottfredson and Hirschi: drug use and crime 212

Key Theories

You should know how to explain the assumptions of these theories and, when applicable, cite related research findings.

Biological theory of deviance
Personality (psychological) theory of deviance
Differential association theory (social learning)
Subcultural deviance
Structural strain
Control theory
Anomie
Labeling

Completion

1. Behavior that does not conform to norms is termed _____.

2. Crime refers to acts of force or fraud undertaken in pursuit of _____.

3. Most crimes are _____ to commit and very _____ in design.

4. Gottfredson and Hirschi proposed that the fundamental psychological feature of those who engage in all varieties of criminal actions is weak _____.

5. Low self-control involves the unwillingness or inability to defer _____.

6. Sutherland proposed the theory of _____, which argued that all behavior is the result of socialization by means of interaction.

7. Subculture deviance can be explained as conflicts over _____.

8. _____ theories attempt to explain deviance on the basis of frustration caused by a person's position in the social structure.

9. _____ argued that deviance is a built-in consequence of stratification.

10. Shapiro has argued that the concept of white-collar crime be replaced with the concept of crimes involving _____.

11. For control theory, the causes of conformity are the _____ between an individual and the group.

12. Types of social bonds include _____, _____, beliefs, and involvements.

13. Anomie is a condition of _____.

14. The Uniform Crime Reports show that the crime rate is _____ in the warmest summer months.

15. Except for rape, crime rates hit a second high point during the month of _____.

16. _____ deviance involves rational calculation and duration.

17. Secondary deviance is a reaction to having been _____.

18. _____ deviation involves actions that cause others to label an individual as deviant.

19. Labeling theorists argue that the _____ a person's status, the less chance he or she will be labeled for deviant behavior.

20. The widespread use of drugs suggests that drugs are not so much a cause of crime as they are a part of the offender _____.

Multiple-Choice

1. Unlike crimes reported in the media, the overwhelming majority of criminal acts:
    a. are very well planned.
    b. are performed incompetently.
    c. result in trivial gains.
    d. b and c above
    e. all of the above

2. Which of the following is/are true about robbery?
    a. Most victims are wealthy and the usual take is over $100.
    b. Most robberies have been extensively planned.
    c. Most of those arrested for robbery are males.
    d. b and c above
    e. all of the above

3. Which of the following is/are true about homicide in the 1990's?
    a. The average victim is killed by someone he or she knows.
    b. Most homicides involve prior planning.
    c. Most killers and their victims are of the same sex and race.
    d. a and c above
    e. all of the above

4. Most criminal actions:
    a. involve short-range choices.
    b. are brief in duration.
    c. are exciting.
    d. b and c above
    e. all of the above

5. Sutherland's theory of differential association attributed deviance to:
    a. biological differences between deviants and nondeviants.
    b. low self-esteem.
    c. strain or frustration caused by poverty.
    d. being labeled deviant.
    e. attachments to others who are deviant.

6.	Gove's research revealed that:
	a.	the arrest rate for violent crimes is highest for ages 30-45.
	b.	females are more likely to be arrested for violent crimes than males.
	c.	the arrest rate for violent crimes drops markedly after age 30.
	d.	a and c
	e.	all of the above

7.	According to Gottfredson and Hirschi, people with low self-control:
	a.	tend to have poor work records.
	b.	tend to be self-centered.
	c.	have an inability or unwillingness to delay gratification.
	d.	all of the above
	e.	none of the above

8.	Subcultural deviance:
	a.	can be explained as conflicts over norms.
	b.	lets us understand that deviance is often a matter of definition.
	c.	explains deviation both among and within subcultural groups.
	d.	a and b
	e.	all of the above

9.	Which of the following men is/are *not* correctly matched with his theory?
	a.	Sutherland: differential association
	b.	Merton: subcultural deviance
	c.	Durkheim: anomie
	d.	Lombroso: born criminals
	e.	b and c

10.	Structural strain theory attempts to explain deviance as a response to:
	a.	a deviant label that stigmatizes a person.
	b.	conflicts over norms.
	c.	a low rate of moral integration.
	d.	deviant attachments.
	e.	none of the above

11. Problems with strain theory include:
    a. studies that have found that a person's social class is barely, if at all, related to committing crimes.
    b. that it seems to predict less deviance than actually occurs.
    c. that it offers no explanation for deviant behavior committed by those in poverty.
    d. a and c
    e. all of the above

12. Bonds between the individual and the group include:
    a. attachments.
    b. beliefs.
    c. investments.
    d. a and c
    e. all of the above

13. Control theory argues that deviant behavior is more likely to occur when:
    a. people have less to gain from deviance than from conformity.
    b. an individual spends a good deal of time and effort on activities that conform to the norms.
    c. the bonds between the individual and the group are weak.
    d. a and b
    e. none of the above

14. The costs we have expended in constructing a satisfactory life and the current and potential flow of rewards coming to us are termed:
    a. investments.
    b. involvements.
    c. attachments.
    d. beliefs.
    e. none of the above

15. Linden and Filmore's study of delinquency found that:
    a. attachments to parents and liking school were negatively correlated with being delinquent.
    b. attachments to parents and liking school were positively correlated with having delinquent friends.
    c. attachments to delinquent peers greatly increase the level of delinquency.
    d. a and c
    e. b and c

16. According to Durkheim, moral communities are characterized by:
    a. high rates of moral integration.
    b. high rates of anomie.
    c. shared beliefs, especially religious beliefs.
    d. a and c
    e. none of the above

17. Recent research on the relationship between crime rates and season have found:
    a. almost no correlation between crime rates and variations in the daily temperature.
    b. crime rates are highest during the hottest summer months.
    c. with the exception of rape, crime rates reach a second peak during the month of December.
    d. b and c above
    e. none of the above

18. In contrast to earlier studies, recent research on the relationship between mental illness and crime has found that discharged mental patients:
    a. pose no more a crime threat than do members of the general population.
    b. are less likely to be arrested than the general public.
    c. are three times as likely as others to be arrested.
    d. seldom commit violent crimes since they are typically receiving drug therapy.
    e. b and d

19. Recent research about the relationship between drug use and crime has found that:
    a. drug use is part of the offender lifestyle.
    b. most offenders began using drugs prior to committing criminal offenses.
    c. most offenders were introduced to drugs after they had begun to offend.
    d. a and b above
    e. a and c above

20. Deviant labels may incline people toward further deviation because:
    a. they limit legitimate occupational opportunities.
    b. being labeled may increase illegitimate economic opportunities.
    c. they can affect self-conceptions.
    d. all of the above
    e. none of the above

Essay

1. A. Explain the theories of differential association, subcultural deviance, and structural strain. (knowledge)
   B. Discuss two criticisms or shortcomings of each of these theories. (comprehension)
   C. Integrate these theories into a more complete explanation of deviance. Use an example. (analysis)

2. A. Explain the labeling theory. (knowledge)
   B. Discuss one criticism or shortcoming of this theory. (comprehension)
   C. Using a fictitious example, apply labeling theory to explain continued deviation. (application)

3. Discuss two biological and/or psychological theories of deviance. Show how research findings did or did not support these theories.

4. Discuss control theory. Be certain to explain in detail the four types of bonds that exist between the group and the individual.

5. What are some of the characteristics of "ordinary crime"? How does the typical offender lifestyle partially explain the high rates of "ordinary crime"?

6. Assume there has been a substantial increase in crime on your campus during the past year and you have volunteered to serve on a task force designed to deal with this problem. How could you use your knowledge of the typical characteristics of ordinary crime to pinpoint potential causes for this increase and offer suggestions to help reduce the crime rate in the future?

Answers

Completion

1.	deviant	11.	social bonds
2.	self-interest	12.	attachments, investments
3.	easy, simple	13.	normlessness
4.	self-control	14.	social integration
5.	gratification	15.	highest
6.	differential association	16.	December
7.	norms	17.	labeled a deviant
8.	Structural strain	18.	primary
9.	Merton	19.	higher
10.	violations of trust	20.	lifestyles

Multiple-Choice

1.	d	11.	a
2.	c	12.	e
3.	c	13.	c
4.	e	14.	a
5.	e	15.	d
6.	c	16.	d
7.	d	17.	d
8.	d	18.	c
9.	b	19.	e
10.	e	20.	d

CHAPTER 8

SOCIAL CONTROL

Overview

Chapter 8 begins with a discussion of social control. It briefly describes mechanisms of informal social control and highlights the research by Asch and Schachter on the effects of the group on the behavior of its members. It then discusses more formal mechanisms of control and efforts of prevention, highlighting the Cambridge-Somerville experiment on delinquency prevention. This chapter also considers Gibbs's theory of deterrence, related research findings, and efforts to reform and resocialize deviants, including the TARP experiment. An appraisal of the current effectiveness of informal and formal mechanisms of social control closes the chapter.

Capsule Summary

Collective efforts to ensure conformity are forms of social control. Most social control is informal and relies on our internalization of the norms, attachments, and power of groups to encourage conformity among their members. Laboratory studies, such as Asch's famous "line experiments" and Schachter's study of group reactions to nonconformity, have empirically demonstrated the power of groups to affect the behavior of their members.

When informal methods fail to produce conformity and the deviance is also illegal or there are legal grounds for intervention, then more formal mechanisms such as police, prisons, and mental hospitals are used. Formal controls have three objectives: (1) to prevent deviance by removing opportunities for it to occur, (2) to deter deviance through the threat of punishment, and (3) to reform or resocialize deviants to discourage them from future deviance.

For a deviant act to occur, there must be an opportunity to commit it. Hence, many programs such as neighborhood-block watches seek to remove these opportunities. It has long been assumed that delinquency has its roots in early socialization. Numerous intervention programs such as the Cambridge-Somerville experiment have attempted to stem this tide; however, such experiments have failed to produce the desired results.

Deterrence is based on the premise that the threat of punishment will discourage deviation. Recent studies conducted by Gibbs and Ehrlich indicate that the threat of punishment may deter deviance provided the punishment is perceived to be swift, certain, and severe. Other studies indicated mixed results, however, and the debate over punishment as deterrence continues. Indeed, the relative ineffectiveness of our current criminal justice system in preventing deviance may stem from the fact that these three criteria are not often met.

Much attention has recently been paid to the therapeutic function of prisons. Attempts have been made to resocialize and reform prisoners in the hope that they would not return to crime upon release. Again, studies such as the TARP experiment do not indicate any significant effect of these efforts, and recidivism rates remain high.

Although recent data may seem pessimistic, social control is effective in many cases. Actually, few people commit crimes; but for those few who do engage in serious deviation, the current criminal justice system has not proven to be very effective.

Key Concepts

You should be able to explain the concepts listed here and be prepared to supply several examples of each.

Social control 217

Informal social control 218

Internalization of norms 218

Formal social control 221

Deterrence 226

Recidivism rate 234

Penitentiary 236

Auburn prison model 236

Resocialization 237

Key Research Studies

You should know both the methodology and the results of these research studies.

Asch: study of group conformity 219

Schachter: group reactions to nonconformity 220

Cabot and others: delinquency prevention by socialization (Cambridge-Somerville experiment) 222

Gibbs: effects of punishment on deterrence 228

Ehrlich: capital punishment and deterrence 230

Lenihan, Rossi, and Berk: financial support of released convicts and recidivism rates (TARP study) 237

Key Theories

Although attempts at prison reform are not theories in the typical sense, they are broader than concepts and do make assumptions, so they are included here.

Opportunity theory

Deterrence theory

Attempts at prison reform

Completion

1. All collective efforts to ensure conformity to the norms are forms of _____.

2. When _____ methods of control fail and more serious acts of deviance occur, _____ methods of social control are activated.

3. Activities by organizations created to ensure conformity to the norms are termed _____.

4. When norms become _____, they become a part of our own beliefs about how we should act.

5. In his famous "line experiments," Asch showed the influence of the group on _____.

6. Formal social control attempts to _____ and to reform or resocialize people.

7. Opportunity theory argues that for a crime to occur, there must be people motivated to commit an offense, suitable targets, and a(n) _____.

8. The Cambridge-Somerville experiment found that the socialization program had _____ effect on delinquency.

9. The use of punishment to discourage people from deviance is termed _____.

10. Gibbs postulated that the more _____, the more certain, and the more _____ the punishment for a crime, the lower the rate at which such a crime will occur.

11. It has been found that what matters is not the actual certainty, swiftness, or severity of punishment but the _____ of these aspects of punishment.

12. Many studies have shown _____ results about capital punishment as deterrence.

13. According to Table 8-3, the most frequently reported crime cleared by arrest in the United States is _____.

14. Studies have found that those who were released from prison after being convicted of violent crimes were slightly _____ likely to have been arrested for a new offense than those convicted of a property offense.

15. According to data from the Bureau of Justice, nearly _____ of people released from prison had been arrested within three years of their release.

16. Approximately _____ as many crimes are committed as are reported.

17. In the 1800s and early 1900s, most prisons were modeled after the _____ design.

18. The proportion of those released from prison who are sentenced to prison again is termed the _____.

19. Efforts to change a person's socialization—to socialize a person over again in hopes of getting him or her to conform to the norms—is termed _____.

20. Whether capital punishment works and whether it is morally justified are _____ questions.

Multiple-Choice

1. Formal social control is attempted through:
    a. preventing deviance.
    b. deterring deviance.
    c. reforming deviants.
    d. a and b
    e. all of the above

2. Asch's experiments found that:
    a. a high proportion of people will conform even in a weak situation.
    b. a low proportion of people will conform even in a weak situation.
    c. the smaller the group, the greater the influence of the group on conformity.
    d. the larger the group, the greater the influence of the group on conformity.
    e. a and d

3. Schachter's study of group conformity found that:
    a. when the paid deviants stuck to their position, they began to receive less attention.
    b. as soon as the paid deviants expressed their views, they began to receive less attention.
    c. group members tended to like the deviant as well as they liked the more conforming members of the group.
    d. all of the above
    e. none of the above

4. Informal social control is attempted by:
    a. internalization of norms.
    b. deterring deviance by threat of punishment.
    c. reforming people.
    d. b and c
    e. all of the above

5. Opportunity theory recognizes that for a crime to occur there must be:
    a. the presence of effective guardians.
    b. suitable targets.
    c. people motivated to commit an offense.
    d. a and c
    e. b and c

6. The Cambridge-Somerville experiment found that:
    a. boys in the experimental group committed fewer delinquent acts than did boys in the control group.
    b. boys in the experimental group committed more delinquent acts than did boys in the control group.
    c. there was no difference in conviction rates between the experimental group and the control group.
    d. the more publicity given an execution, the lower the homicide rate.
    e. the smaller the group, the greater its influence on conformity.

7. Experiments in delinquency prevention have generally:
    a. been highly successful in preventing delinquency.
    b. been failures in preventing delinquency.
    c. succeeded in changing the life circumstances of the children.
    d. a and c
    e. none of the above

8. Gibbs postulated that the more _____ the punishment for a crime, the lower the rate at which such a crime will occur.
    a. severe
    b. certain
    c. swift
    d. all of the above

9. Gibbs's theory of deterrence:
    a. predicts that severe sentences will not effectively deter crimes if people realize that they have little chance of being caught.
    b. can be applied to all deviant acts.
    c. was not supported by empirical research.
    d. a and c
    e. all of the above

10. Results of studies of deterrence have found that:
    a. capital punishment always deters crime.
    b. capital punishment almost always deters crime.
    c. capital punishment has no deterrent effect on homicide.
    d. no one knows for certain whether capital punishment deters homicide.
    e. none of the above

11. The use of punishment to discourage people from committing deviance is termed:
    a. prevention.
    b. deterrence.
    c. resocialization.
    d. revenge.
    e. reform.

12. Problems surrounding the subject of deterrence include:
    a. the difficulty of proving that deviant acts that do not happen have, in fact, been deterred.
    b. the degree to which fear of punishment explains acts of conformity.
    c. that some people—for example, those in prison—were not deterred.
    d. all of the above
    e. none of the above

13. The case against capital punishment has been argued on the basis of:
    a. religion.
    b. racial discrimination.
    c. conflicting research results.
    d. a and b
    e. all of the above

14. The relatively high rates of arrest for the crime of murder are achieved because:
    a. most murders are acts of sudden passion rather than the result of extensive planning.
    b. most murders are committed by a friend or relative of the victim.
    c. the murderer is often apprehended at the scene of the crime.
    d. all of the above
    e. none of the above

15. Table 8-3 indicates the crime *most likely* to be solved in the United States is:
    a. assault.
    b. rape.
    c. robbery.
    d. homicide.
    e. a and b

16. Punishment for those who commit crimes in the United States is:
    a. very certain.
    b. very swift.
    c. very severe.
    d. all of the above
    e. none of the above

17. The first group to experiment with the use of a penitentiary was:
    a. the ancient Greeks under Plato.
    b. the Quakers in Pennsylvania.
    c. the British during the 1700s.
    d. Cabot's Cambridge-Somerville experiment.
    e. none of the above

18. Reform efforts in prison are probably hampered by:
    a. a further weakening of an inmate's attachments with conventional people.
    b. the stigma associated with being an ex-convict, which hinders the formation of attachments with conventional people.
    c. new attachments made to other deviants while a person is in prison.
    d. a and b
    e. all of the above

19. In the TARP experiment, the recidivism rate for the control group was _____ than the experimental group's.
    a. much lower
    b. much higher
    c. slightly lower
    d. slightly higher
    e. the same

20. Which of the following statements is/are true?
   a   If crime can be reduced by overhauling the criminal justice system, then we could expect truly dramatic changes.
   b.   A great deal of our conformity is rooted in informal social control.
   c.   Social control does not appear to work because more than 13 million crimes are reported each year.
   d.   all of the above
   e.   none of the above

Essay

1.   A.   Name the three criteria that must be present if punishment is to serve as a deterrent. (Gibbs) (knowledge)
     B.   Explain Gibbs's theory of deterrence. (comprehension)
     C.   What does Gibbs's theory imply about our current criminal justice system? (application)

2.   A.   Describe either Asch's or Schachter's study of group conformity. (comprehension)
     B.   Show how these results might affect real-world (rather than laboratory) social interaction. (application)
     C.   Compare and contrast these studies in terms of methodology and results. (analysis)

3.   Discuss the Cambridge-Somerville experiment. What do the results of this and other intervention programs imply about delinquency prevention?

4.   Explain the following statement: "The question of whether capital punishment deters homicide is entirely distinct from the moral debate over the use of capital punishment."

5.   Briefly trace the history of prisons in the United States. Explain why contemporary prisons largely fail in their mission to reform inmates.

6.   Last semester you noticed that several students regularly sat together in class and appeared to cheat on exams. The proctor seemed not to notice or call attention to this behavior. How would you explain the cheating using your knowledge of formal and informal social control?

Assume you heard that two of these students were caught cheating in another class and received disciplinary action. This semester you notice these same students sit together but there seems no evidence of cheating. How might Gibb's theory of deterrence be applied here? What other elements of formal and informal social control might also be operating?

Answers

Completion

1.	social control	11.	perceptions
2.	informal, formal	12.	mixed
3.	formal methods of social control	13.	homicide (murder)
4.	internalized	14.	less
5.	an individual's conformity	15.	two-thirds
6.	prevent deviance	16.	twice
7.	absence of effective guardians	17.	Auburn
8.	no	18.	recidivism rate
9.	deterrence	19.	resocialization
10.	rapid, severe	20.	unrelated

Multiple-Choice

1.	e	11.	b
2.	a	12.	d
3.	a	13.	e
4.	a	14.	d
5.	e	15.	d
6.	c	16.	e
7.	b	17.	b
8.	d	18.	e
9.	a	19.	e
10.	d	20.	b

CHAPTERS 5 TO 8

REVIEW AND SPECIAL PROJECT

Review

Chapters 5 through 8 have discussed the relationship between the individual and the group, as well as the biological and cultural basis for behavior. They emphasized informal and formal social control and described the various theories of deviation. You may wish to apply your knowledge of this material to the following project.

Special Project

Conduct an interview with someone who has committed an act that would qualify as a serious deviance. This need not be an illegal act; it could be behavior that, if discovered, would have resulted in trouble with the police, school authorities, or employers. The subject need not have been caught. It will probably not be difficult to locate such a person. Remember that what may be conforming in one subculture may be deviant in the general culture.

Be certain to secure your subject's consent; explain to him or her the nature and purpose of your interview and respect your subject's confidentiality by allowing your subject to remain anonymous if he or she asks to be unnamed.

During your interview, try to obtain as much information as you can about the following:

1.   The subject's situation at the time of the deviance: age, employment, family situation, personal problems, and so on
2.   The subject's background: education, religion, family, and social class
3.   The actual situation in which the deviant act occurred: alone or with others, spur of the moment or planned
4.   The purpose of the deviation: fun, material gain, the venting of anger, and so on
5.   Has the deviation been repeated? If so, why? If not, why not?
6.   Was the subject caught? If so, what happened?

After you have conducted your interview, attempt to explain this deviation by using some of the theories you have studied. Focus on the mechanisms of social control that may or may not have been effective. Does this case seem to fit a particular theory, or does your explanation entail components of several theories? Do any theories fail to fit this case at all? Attempt to develop your own theory of deviance that is applicable to this situation.

CHAPTER 9

CONCEPTS AND THEORIES OF STRATIFICATION

Overview

This largely theoretical chapter describes modern social theories of stratification. It opens with a discussion of Marx's two-class model and then discusses Weber's three-dimensional approach to social class. It describes Lenski's theory of status inconsistency and cites related research findings. It distinguishes between ascribed and achieved statuses and differentiates between structural and exchange mobility.

Chapter 9 then turns its attention to Marx's utopian classless society and Dahrendorf's critique of Marx. After Mosca's theory of the inevitability of stratification is summarized, the chapter examines the functionalist, social evolutionary, and conflict approaches to stratification. Throughout the discussion of these theoretical approaches, an example of a "toy society" illustrates the various concepts and assumptions. This chapter offers a more in-depth explanation of many of the concepts introduced in Chapter 2 and sets the stage for many of the chapters that follow.

Capsule Summary

Social stratification, the unequal distribution of rewards in a society, has long been the subject of considerable interest. Divisions of wealth and rank within societies are termed social classes.

Marx, who developed the first modern social theory of stratification, viewed class from the economic dimension. To Marx, class membership was based on one's relationship to the means of production. Those who own the means of production were termed bourgeoisie, and those who work it were termed proletariat. Class membership was defined by both one's material position in society and one's class consciousness. Marx was a utopian who argued for a classless society based on the abolition of private property. In his critique of Marx, Dahrendorf argued that communistic societies were only classless by definition. Stratification in these societies was based on control rather than ownership of the means of production.

Weber argued for a broader definition of class, with class membership determined by three dimensions: class (also termed property), status (also called prestige), and power (also called party). People disproportionately high (or low) on one dimension are termed status inconsistent. Research has found that status-inconsistent people often favor politically liberal causes.

Status may be based on achievement or ascription. Caste systems rely predominantly on ascription, whereas industrialized societies rely more heavily on achievement. Social mobility, which results from changes in the distribution of statuses, is termed structural mobility. Mobility that is not structural is termed exchange mobility.

Mosca argued that stratification is the inevitable result of political organization, which fosters inequalities in power and allows those with greater power to exploit others for material advantage.

The functional, social evolutionary, and conflict theories differ in their assumptions, yet all assume that some degree of stratification is inevitable. Functionalists argue that the stratification system ensures that functionally important positions will be filled because those occupying such positions will be highly rewarded. The concepts of functional importance and the principle of replaceability are central to the functionalist explanation. Social evolutionists argue that cultural accumulation results in a division of labor and specialization. This in turn leads to stratification. Conflict theorists argue that societies are even more stratified than necessary because those in powerful positions use their power to exploit the less powerful. Labor unions and professions, for example, use their power to control replaceability and ensure their own continuation of power.

Key Concepts

You should be able to explain the concepts listed here and be able to cite several examples of each.

Social mobility 245	Power (party) 251
Class 246	Status characteristics 251
Bourgeoisie 247	Status inconsistency 251
Proletariat 247	Social mobility 252
Means of production 247	Caste system 252
Lumpenproletariat 247	Structural mobility 254
Class consciousness 247	Exchange mobility 254
False consciousness 248	Utopian 257
Property (class) 249	Anarchists 257
Prestige (status) 250	Replaceability 260

Key Research Studies

You should be familiar with both the methodology and the results of the tests of the theory of status inconsistency by these researchers.

Gary Marx 252
Cohn 252
Baltzell 252

Key Theories

You should be prepared to explain the assumptions of these theories and, when applicable, cite related research findings.

      Marx: stratification
      Weber: stratification
      Lenski: status inconsistency theory
      Dahrendorf: critique of Marx
      Mosca: stratification
      Davis and Moore: functionalist viewpoint
      Evolutionary perspective
      Conflict perspective

Completion

1. Upward or downward movement by individuals or groups within a stratification system is termed _____.

2. Divisions of rank and wealth within societies are termed _____.

3. Marx termed the class of people who work the means of production the _____.

4. Marx termed the very bottom of society's stratification system the _____.

5. Marx termed the tendency for workers to believe they had common interests with the ruling class _____.

6. Marx's definition of class is determined only by the _____ dimension.

7. Weber termed the three dimensions of stratification _____, _____, and _____.

8. Weber considered groups of people with similar life chances as determined by their economic position in social _____.

9. The ability to get one's way despite the resistance of others is _____.

10. Modern social scientists consider the "Three P's" of stratification to be _____, _____, and _____.

11. Certain individual or group traits that determine status are termed _____.

12. Persons or groups who hold different ranks on each of the three dimensions are termed _____.

13. Table 9-1 shows that the essential principle of the _____ theory is widely accepted.

14. _____ status is based on merit, while _____ status is a position based on who you are.

15. Social mobility that results from changes in the distribution of statuses in society is called _____.

16. A(n) _____ constructs plans for an ideal society.

17. _____ argued that stratification cannot be avoided because it is an inescapable feature of collective life.

18. A position is of high functional importance to a society to the degree that either the _____ or its _____ are hard to replace.

19. _____ theories focus on how stratification systems are subject to distortion.

20. _____ argue that the accumulation of culture inevitably leads to a division of labor and therefore to stratification.

Multiple-Choice

1.   Marx termed those who work the means of production:
   a.   the bourgeoisie.
   b.   the proletariat.
   c.   the lumpenproletariat.
   d.   the middle class.
   e.   none of the above

2.   Which of the following groups was not incorporated into Marx's class system?
   a.   the owners of the means of production
   b.   farmers and peasants
   c.   lumpenproletariat
   d.   b and c
   e.   all of the above

3.   Marx defined the two classes on the basis of:
   a.   their material position in society.
   b.   prestige.
   c.   power.
   d.   a and c
   e.   all of the above

4.   What Weber called *class* modern social scientists refer to as:
   a.   prestige.
   b.   status.
   c.   property.
   d.   power.
   e.   none of the above

5.   The ability to get one's way despite the resistance of others is termed:
   a.   property.
   b.   class.
   c.   power.
   d.   prestige.
   e.   none of the above

6.   When famous sports stars endorse a commercial product, they are exchanging their _____ for economic advantage.
   a.   property
   b.   prestige
   c.   power
   d.   class
   e.   position

7.   Which of the following would be *most* likely to experience status inconsistency?
   a.   a white male lawyer
   b.   a black engineer with a doctorate
   c.   a female physician
   d.   b and c
   e.   a and c

8.   Research in the area of status inconsistency has found that:
   a.   upper-status African-Americans are more radical than lower-status African-Americans.
   b.   Jewish bankers and industrialists have a record of voting for conservative parties.
   c.   wealthy and powerful American Catholics have a strong preference for the Republican party.
   d.   all of the above
   e.   none of the above

9. In opting for political responses, people suffering from status inconsistency:
   a. usually blame themselves.
   b. usually blame others—that is, the "system."
   c. are more likely to blame themselves if their inconsistent status is a result of group characteristics such as race, religion, or ethnicity.
   d. a and c
   e. none of the above

10. The notion of status inconsistency is possible:
    a. only if we accept Weber's view of multiple bases for ranks in society.
    b. only if we accept Marx's view of multiple bases for ranks in society.
    c. if we accept either Marx or Weber's multidimensional view.
    d. only if we accept only one basis for rank as Marx does.
    e. only if we accept one basis for rank as Weber does.

11. When a society uses ascriptive status rules, people may be placed in status positions based on _____
    a. place of birth.
    b. family background.
    c. sex.
    d. b and c
    e. all of the above

12. In a caste system:
    a. status is based entirely on ascription.
    b. ascription is the overwhelming basis for status.
    c. status is based entirely on achievement.
    d. achievement is the overwhelming basis for status.
    e. none of the above

13. Exchange mobility:
    a. is common when the ascriptive status rule operates.
    b. is very uncommon when status is based on achievement.
    c. is very common regardless of the rules governing status.
    d. a and b
    e. none of the above

14. Marx argued that to achieve a classless society:
    a. the proletariat should own the means of production.
    b. the state should be abolished.
    c. private ownership of the means of production should be abolished.
    d. the bourgeoisie should own the means of production.
    e. none of the above

15. Mosca argued that:
    a. human societies cannot exist without political organization.
    b. whenever there is political organization, there must be inequalities in power.
    c. a classless society is possible if private ownership is abolished.
    d. a and b
    e. b and c

16. The functionalist view of stratification argues that:
    a. positions in society differ in the degree to which they are functionally important.
    b. some positions are inherently more difficult to fill than others.
    c. to ensure an adequate supply of people to fill important positions, it is necessary to attach higher rewards to those positions.
    d. all of the above
    e. none of the above

17. Which of the following theorists is/are correctly paired with his view or theory of stratification?
    a. Lenski: social evolutionary
    b. Davis and Moore: conflict
    c. Marx: functionalist theory
    d. all of the above
    e. none of the above

18. A position is of high functional importance when:
    a. the position itself is hard to replace.
    b. the occupants of the position are hard to replace.
    c. its functions can be performed by people in other positions.
    d. a and b
    e. all of the above

19. _____ theory is based on the premise that the accumulation of culture results in cultural specialization, which in turn results in stratification.
    a. Functionalist
    b. Social evolutionary
    c. Conflict
    d. Marxist
    e. Weberian

20. Most modern conflict theorists:
    a. argue that stratification is unavoidable.
    b. assume that people who are in a position to exploit others will do so.
    c. argue that societies are more stratified than necessary.
    d. b and c
    e. all of the above

Essay

1. A. List the three theories of stratification. (knowledge)
   B. Explain the three theories of stratification. (comprehension)
   C. Compare and contrast two of these theories. (analysis)

2. A. Define the principle of replaceability. (knowledge)
   B. Give examples of the politics of replaceability from labor unions and the professions. (comprehension)
   C. Apply this principle to the "toy society." (application)

3. Contrast Marx and Weber on the subject of social class.

4. Discuss the concept of status inconsistency. Cite research findings on this topic.

5. Discuss Marx on the classless society and explain Dahrendorf's critique.

6. Much of the research on status inconsistency and political liberalism was conducted during the 1960's when women and minorities were relatively rare in most professions. While complete equality has not yet been achieved, the last three decades have seen significant changes in this area. How do you feel these changes may have affected the concept of status inconsistency and its ramifications? Do you feel the relationship between status inconsistency and political liberalism is stronger or weaker than it was in the 1960's? How might you use the results of recent elections to test your assumptions?

Answers

Completion

1. social mobility
2. social classes
3. proletariat
4. lumpenproletariat
5. false consciousness
6. economic
7. class, status, power
8. classes
9. power (or party)
10. property, prestige, power
11. status characteristics
12. status inconsistent
13. functionalist
14. Achieved, ascribed
15. structural mobility
16. utopian
17. Mosca
18. position itself, occupants
19. Conflict
20. Evolutionary theorists

Multiple-Choice

1. b
2. d
3. a
4. c
5. c
6. b
7. d
8. a
9. b
10. a
11. e
12. b
13. e
14. c
15. d
16. d
17. a
18. d
19. b
20. e

CHAPTER 10

COMPARING SYSTEMS OF STRATIFICATION

Overview

Chapter 10 contrasts the stratification systems of hunting and gathering, agrarian, and industrialized societies. It starts with a description of hunting and gathering societies and then describes the changes brought about by the advent of agrarian societies. It also describes the stratification systems of agrarian societies and examines the differences between the elite and the masses. The chapter then discusses the changes in the stratification system brought about by industrialization, focusing on research findings about mobility in industrialized nations. "A Closer View" provides an in-depth look at several studies on status attainment. Many of the concepts first introduced in Chapters 2 and 9 are applied to the various stratification systems. The chapter ends with Special Topic 2, the aspects of income inequalities.

Capsule Summary

Stratification exists in all societies, even the simplest hunting and gathering societies composed of small bands of people who wander in search of food. They are the least stratified of all societies and have few possessions, no full-time leaders, and little role specialization. Stratification is typically based on age and sex, although within the sexes it is often based on achieved characteristics.

As societies became more complex, they became more stratified. Agrarian societies fostered the rise of specialization, personal property, government, cities, and slavery. These societies were highly stratified, with large gaps separating the elite from the masses. Agrarian societies were dominated by the military and were often in a chronic state of warfare.

Industrialization changed the stratification system. Societies became less stratified. The gap between the top and bottom decreased as the middle classes expanded and jobs for unskilled labor began to disappear. As more skill and training were required for jobs, positions and their occupants became less replaceable; workers thus became more powerful and better able to resist coercion. The rise of democracy and industrialization went hand in hand.

Research has found high rates of structural mobility in industrialized nations. Compared with European nations, the United States has higher rates of long-distance mobility. Recent research by Blau and Duncan, Porter, Cohen and Tyree, Mare and Tzeng, and Hout has focused on status attainment in the United States and Canada and the application of the status attainment model to mobility in these nations. The results of research have yielded some surprising results. Although earlier studies of mobility focused only on the male, more recent studies have also focused on female mobility and status attainment.

Key Concepts

You should be able to explain the following concepts and be prepared to provide several examples of each.

Hunting and gathering society 265	Industrial society 275
Agrarian society 268	Industrialization 275
Specialization 268	Structural mobility 278
Urbanization 268	Long-distance mobility 279
	Status attainment model 280

Key Research Studies

You should be familiar with both the methodology and the results of the research studies cited here.

Lipset and Bendix: comparative social mobility 278

Cross-cultural research on long-distance mobility 279

Blau and Duncan: status attainment model 280

Cohen and Tyree: comparison of mobility of those from poor versus nonpoor
        backgrounds 281

Mare and Tzeng: advantage of older parents 282

Porter et al.: status attainment in Canada 282

Hout: trends in status attainment 285

World Values Survey: income and life-style (Special Topic 2) 287

Key Theories

You should be able to explain the assumptions of these theories as they apply to stratification in industrial societies and when applicable, cite related research findings.

Functionalism

Conflict theory

In addition to these theories, you should be *thoroughly* familiar with stratification systems of the following:

Hunting and gathering societies

Agrarian societies

Industrial societies

Completion

1. The simplest type of society is the _____ society.

2. Recent archeological discoveries in Africa indicate that humans existed more than _____ years ago.

3. The primary bases for stratification in hunting and gathering societies are age and _____.

4. Within age and sex groups, simple societies are not very _____.

5. With the invention of plows and effective animal harnesses, _____ societies appeared.

6. The advent of agriculture made possible _____ and _____.

7. _____ societies lived in a chronic state of warfare.

8. Because agrarian societies were based on _____, they tended to be expansionistic.

9. The most stratified societies are typically _____ societies.

10. Industrial societies are not based on getting people to work harder but on getting them to _____.

11. Using technology to make work much more productive is termed _____.

12. As positions become _____ replaceable, their relative rewards increase.

13. As a consequence of industrialization, the average worker is more _____ and thus more able to resist coercion.

14. Compared to agrarian societies, industrial societies are _____ stratified.

15. Bendix and Lipset observed that in industrial societies there is a great deal of _____ mobility.

16. _____ mobility involves high upward or downward shifts in status.

17. Blau and Duncan found a very strong correlation between _____ and occupational status.

18. Studies of status attainment in Canada have found that it is very _____ to that in the United States.

19. Cohen and Tyree have found the main determinant of the family income of persons from all backgrounds is _____.

20. Hout's study of trends in status attainment found that while structural mobility has been declining there has been an increase in _____ mobility.

Multiple-Choice

1. The simplest form of society is the _____ society.
   a. gardening
   b. hunting and gathering
   c. agrarian
   d. industrial
   e. herding

2. The least stratified societies are:
   a. agrarian societies.
   b. hunting and gathering societies.
   c. industrialized societies.
   d. herding societies.
   e. horticultural societies.

3. In a hunting and gathering society:
   a. there is extensive role specialization.
   b. power is based on land ownership.
   c. stratification is generally based on sex and age.
   d. a and c
   e. all of the above

4. The rule seems to be the _____ a human society is, the less stratified it is.
   a. smaller
   b. poorer
   c. more secure
   d. a and b
   e. all of the above

5. Improved agriculture produced:
   a. warfare.
   b. cities.
   c. surplus.
   d. b and c
   e. all of the above

6. Slavery is most likely to be found in:
   a. hunting and gathering societies.
   b. agrarian societies.
   c. industrialized societies.
   d. a and b
   e. all of the above

7. The ability to produce surplus food may lead to:
   a. humans becoming property.
   b. a more extensive division of labor.
   c. government.
   d. all of the above
   e. none of the above

8. In an agrarian society, the masses and the elite may differ in:
   a. language.
   b. ethnic background.
   c. the amount of time devoted to leisure activities.
   d. a and c
   e. all of the above

9. The most highly stratified societies tend to be:
   a. agrarian.
   b. herding.
   c. gardening.
   d. industrialized.
   e. b and c

10. Changes in the stratification system brought about by industrialization include:
    a. a widening of the gap between the masses and the elite.
    b. positions becoming more replaceable.
    c. a decrease in the power of the average worker, who is less able to resist coercion.
    d. all of the above
    e. none of the above

11. Which of the following statements is/are true?
   a. As positions become less replaceable, their relative rewards decrease.
   b. To the extent that occupational positions require education and training, they are less replaceable.
   c. In the long run, industrialization has made the average worker more replaceable.
   d. a and c
   e. all of the above

12. As a consequence of industrialization:
   a. people "work smarter."
   b. the average worker has become less replaceable.
   c. the training necessary to do most jobs has enabled workers to demand a higher level of reward.
   d. all of the above
   e. none of the above

13. Differences between industrialized and agrarian societies indicate that:
   a. industrial societies are more stratified.
   b. industrial societies are less stratified.
   c. the gap between the poor and the wealthy is greater in an industrial society.
   d. a and c
   e. b and c

14. Cohen and Tyree found that:
   a. education has less importance for status attainment for people from poor homes than for other people.
   b. education has greater importance for status attainment for people from poor homes than for other people.
   c. the main determinant of the family income of persons from all backgrounds is marital status.
   d. a and c
   e. b and c

15. Huge upward and downward shifts in status are termed _____ mobility.
   a. exchange
   b. structural
   c. long-distance
   d. vertical
   e. horizontal

16. Blau and Duncan (status attainment model) found that:
   a. it is better to start at the top than at the bottom of the stratification system.
   b. the primary mechanism linking the occupational status of fathers and sons is education.
   c. family background is a more important influence on status attainment than is education.
   d. a and b
   e. a and c

17. Studies of status attainment in Canada have found that:
   a. Canada's rate of social mobility is almost identical to that of the United States.
   b. opportunities for occupational advancement in Canada were much more restricted than in most industrialized nations.
   c. ethnicity plays a major role in status attainment.
   d. all of the above
   e. b and c

18. Studies of female status attainment in Canada have shown that:
   a. native-born Canadian women with full-time jobs come from lower-status family backgrounds than do their male counterparts.
   b. the average native-born Canadian woman has a higher-status occupation than do similar males.
   c. the correlations between women's occupational prestige and their fathers' occupational prestige are much lower than for men.
   d. b and c
   e. all of the above

19. Cross-cultural research on long-distance mobility has found that:
   a. Americans are much more likely than people in other industrial nations to object to current income differences.
   b. Americans are more likely than people in other industrial nations to rate "hard work" as very important in getting ahead.
   c. most Americans think they need to come from a wealthy family in order to get ahead.
   d. all of the above
   e. none of the above

20. Studies of trends in status attainment have found that:
   a. structural mobility is on the decline.
   b. exchange mobility is on the decline.
   c. structural mobility is on the increase.
   d. the stratification system in the United States has become less open.
   e. both structural and exchange mobility rates have remained about the same.

Essay

1. A. List the three types of societies. (knowledge)
   B. Explain the stratification systems in each of these societies. (comprehension)
   C. Contrast the stratification systems of hunting and gathering societies with those of agrarian societies. (analysis)

2. A. Explain the stratification systems in agrarian and industrialized societies. (comprehension)
   B. Show how industrialization changed the stratification system. (application)
   C. Contrast the stratification systems of these two societies. (analysis)

3. Discuss some of the research findings on social mobility.

4. Explain the following statement: "The smaller, the poorer, and the less secure society is, the less it is stratified."

5. Discuss some recent changes in trends in status attainment.

6. If the information is available to you, you might wish to investigate your own family background for evidence of structural and/or exchange mobility. Do you find your family history is consistent with the typical trends in status attainment discussed in this chapter? Do you notice any differences between your situation and the typical trends? If so, what other factors may help account for these differences?

Answers

Completion

1. hunting and gathering
2. 3 million
3. sex
4. stratified
5. agrarian
6. specialization, cities
7. Agrarian
8. military rule
9. agrarian
10. work smarter
11. industrialization
12. less
13. powerful
14. less
15. structural
16. Long-distance
17. education
18. similar
19. marital status
20. exchange

Multiple-Choice

1. b
2. b
3. c
4. d
5. e
6. b
7. d
8. e
9. a
10. e
11. b
12. d
13. b
14. e
15. c
16. d
17. a
18. d
19. b
20. a

CHAPTER 11

INTERGROUP CONFLICT:
RACIAL AND ETHNIC INEQUALITY

Overview

Chapter 11 opens with a discussion of intergroup conflict and mechanisms for reducing these conflicts. It explains early theories such as authoritarianism, which argued that conflict results from prejudice. Then it focuses on more recent explanations such as Allport's contact theory, which argued that prejudice results from social inequality and competition. Using examples, the chapter shows how inequality and economic competition may lead to prejudice. This chapter also examines slavery and its aftermath and discusses Myrdal's *An American Dilemma*. The decline of prejudice is then discussed with examples such as the case of Japanese immigrants in the United States and Canada. The chapter then turns its attention to Hispanic-Americans. It also considers today's minorities and analyzes their plight and recent progress from the perspective of "new immigrants." The chapter closes with a discussion of the importance of enclave economies, integration patterns, and barriers to African-American progress. Special Topic 3, which focuses on minority participation in sports and entertainment as a mechanism for overcoming discrimination, is included at the end.

Capsule Summary

Prejudice is often the result of intergroup conflict and status inequality. Intergroup conflict may include both racial and ethnic groups. This conflict may be resolved through assimilation, accommodation (and hence pluralism), extermination or expulsion of the weaker group, or the imposition of a caste system. Earlier theories such as the authoritarian personality theory argued that prejudice caused intergroup conflict. More recent theories argued that the reverse is correct.

Allport's contact theory argued that contact between groups will reduce prejudice if the two groups meet on the basis of equal status and pursue common goals. Research has supported this claim. Conditions imposed by slavery and its aftermath (sometimes termed "the American dilemma") illustrate the reverse of this theory. Inequality, economic competition, and the imposition of a caste system can increase prejudice. Minorities do not always occupy the lowest status. Middleman minorities frequently are a buffer between the highest and lowest classes and are often the targets of their frustration.

Historically, many minorities have achieved upward mobility through the mechanisms of geographic concentration, internal economic development and occupational specialization, the development of a middle class, and the creation of enclave economies. Japanese-Americans and Japanese-Canadians are prime examples.

Contrary to popular opinion, Hispanic-Americans do not constitute a single ethnic group. Rather, the social and cultural differences among people of Mexican, Puerto Rican, and Cuban heritage indicate that these groups each constitute distinct ethnic groups.

Today's minorities—African-Americans, Native Americans, and Eskimos, for example—differ from other minorities in that they have resided in the United States and Canada for many generations. Still, they have been treated almost as a caste apart from mainstream society. For this reason, many sociologists view them as recent immigrants and analyze their recent progress from that standpoint. Until recently, African-Americans, for example, have been concentrated in the rural South, where they received little education or training necessary for upward mobility in an urban, industrialized society. They have been further hampered by their visibility, large population, lack of a "homeland," and the legacies of slavery. Prejudice still exists, although recent gains made by these minorities are reflected in integration patterns in residence, church attendance, and an increase in the number of interracial friendships and interracial marriages. Research seems to indicate that prejudice is indeed declining.

Key Concepts

You should be able to explain the concepts listed here, as well as cite several examples of each.

Intergroup conflict 297	Expulsion 300
Race 297	Segregation 300
Ethnic group 297	Prejudice 300
Cultural (ethnic) pluralism 298	Authoritarian personality 301
Assimilation 298	"An American Dilemma" 306
Accommodation 300	Cultural division of labor 314
Extermination 300	Middleman minorities 314

Key Research Studies

You should be familiar with both the methodology and the results of these research studies.

Various studies of the relationship between prejudice and other variables 301
Sherif and Sherif: artificial inducement of prejudice 303
de la Garza et al.: Hispanic-American perspectives on American politics 330
Firebaugh and Davis: *The Decline of Prejudice* 338

Key Theories

You should be prepared to explain the assumptions of the following theories and, when applicable, cite related research findings. Blalock's theory appears in Special Topic 3.

Theories of prejudice: authoritarian personality, contact theory, status inequality

Bonaduch's model: labor and minorities

Three mechanisms that foster upward mobility

Enclave economy theory

African-Americans as recent immigrants

Blalock's explanation of minorities' success in sports (Special Topic 3)

In addition, you should be *thoroughly* familiar with the experiences of the various minorities discussed in this chapter.

Completion

1. A human group with some observable common biological feature is termed a(n) _____.

2. Ethnic groups are groups whose _____ differ.

3. For a long time it was believed that intergroup conflicts in North America would be resolved through _____.

4. When intergroup conflict ends through _____, the result is ethnic or cultural pluralism.

5. Besides accommodation and assimilation, the possible outcomes of intergroup conflict include _____, _____, and _____.

6. Until very recently, most social scientists regarded _____ as the cause of intergroup conflict.

7. Adorno argued that prejudice was the result of a(n) _____ personality type.

8. Contact theory is most closely associated with _____.

9. Allport argued that prejudice will _____ if two groups are engaged in competition.

10. The more highly _____ a person is and the higher a person's _____, the less likely he or she will be prejudiced against other racial and ethnic groups.

11. Myrdal's *An American Dilemma* dealt with the contradiction between _____ and _____.

12. There is no greater status inequality than that between _____ and _____.

13. Lieberson concluded that fear of African-Americans as _____ is the real cause of racial stereotypes.

14. According to Bonaduch, factors that lead subordinate groups to work for substandard wages include a low standard of living, a lack of political power, _____, and _____.

15. A minority that serves as both a link and a buffer between the upper and lower classes is referred to as a(n) _____ minority.

16. Sociologists today consider the experience of African-Americans in the United States to be similar to that of _____.

17. Research by de la Garza et al. found that while most subjects spoke both English and Spanish, most preferred to be interviewed in _____.

18. The degree to which a racial or ethnic group can be recognized is termed _____.

19. Firebaugh and Davis found that both _____ and generational replacement have made major contributions to the decline of prejudice.

20. Barriers to African-Americans' progress include the legacies of slavery, visibility, _____, and _____.

Multiple-Choice

1.   Races may differ on the basis of:
        a.   skin color.
        b.   eyelid shape.
        c.   blood type.
        d.   a and b
        e.   all of the above

2. Possible outcomes of intergroup conflict include:
   a. accommodation.
   b. expulsion of the weaker group.
   c. the imposition of a caste system.
   d. a and b
   e. all of the above

3. Ethnic groups are:
   a. groups whose cultural heritage differs from other groups within the same society.
   b. "involuntary" groups because people do not choose to join them.
   c. groups that share observable biological differences.
   d. a and b
   e. a and c

4. When intergroup conflict ends through accommodation, the result is:
   a. ethnic or cultural pluralism.
   b. assimilation.
   c. expulsion of the weaker group.
   d. all of the above
   e. none of the above

5. The theory that some people are so oversocialized that they accept only their group's norms and values argues that prejudice is the result of:
   a. intergroup conflict.
   b. status inequality.
   c. competitive contact with others.
   d. the authoritarian personality.
   e. none of the above

6. Contact theory is most closely associated with:
   a. Gordon Allport.
   b. Gunnar Myrdal.
   c. Muzafer Sherif.
   d. H. M. Blalock.
   e. none of the above

7. According to contact theory, prejudice will *intensify* when the groups:
   a. possess equal status in the situation.
   b. cooperate to pursue common goals.
   c. are engaged in competition.
   d. a and b
   e. all of the above

8.   Which of the following is/are true about the history of slavery?
       a.   Until quite recently slavery was nearly universal.
       b.   Slavery has always been associated with racial minorities; at no time in
            history have whites been used as a source of slaves.
       c.   Virtually all societies world-wide had abolished slavery by the end of the
            eighteenth century.
       d.   a and b above
       e.   all of the above

9.   Today most sociologists view status inequality as:
       a.   the cause of prejudice but the result of discrimination.
       b.   the cause of, not the result of, prejudice and discrimination.
       c.   the result of prejudice and discrimination.
       d.   the cause of discrimination and the result of prejudice.
       e.   none of the above

10.   Lieberson believes that the real *cause* of racial stereotypes is:
       a.   fear of black-and-white intermarriage.
       b.   differences in skin color.
       c.   prejudice and discrimination.
       d.   fear of African-Americans as economic competitors.
       e.   none of the above

11.   Bonaduch argued that factors that often cause or require members of
       subordinate groups to work for substandard wages include:
       a.   a low standard of living.
       b.   a lack of information.
       c.   a lack of political power.
       d.   all of the above
       e.   a and c

12.   Strategies used to prevent a subordinate group from economically competing
       with the dominant group include:
       a.   exclusion.
       b.   the establishment of a caste system.
       c.   accommodation.
       d.   a and b
       e.   all of the above

13. Middleman minorities:
    a. may serve as a link between the upper and lower classes.
    b. may serve as a buffer between the upper and lower classes.
    c. may defuse potential class conflicts by becoming the focus of frustration and anger.
    d. b and c
    e. all of the above

14. Compared with other Hispanic-American groups, the rapid upward mobility of Cuban-Americans can partly be attributed to their:
    a. older median age.
    b. higher educational level.
    c. greater fluency in English.
    d. a and b
    e. all of the above

15. Mechanisms by which minorities have achieved upward mobility include:
    a. occupational generalization.
    b. geographical concentration.
    c. the development of a middle class.
    d. b and c
    e. all of the above

16. Internal economic development by minorities is often enhanced by:
    a. the founding of their own financial institutions.
    b. the development of an enclave economy.
    c. buying outside their own community.
    d. a and b
    e. all of the above

17. Until very recently, African-Americans in the United States:
    a. lived predominantly in the South.
    b. lived mainly in urban areas.
    c. had a low level of education.
    d. a and c
    e. all of the above

18. Barriers to African-American progress and upward mobility include:
    a. the legacies of slavery.
    b. the lack of a "homeland."
    c. visibility.
    d. all of the above
    e. none of the above

19. According to research by Firebaugh and Davis:
    a. prejudice against African-Americans declined substantially between 1972 and 1984.
    b. prejudice has declined more rapidly in the North than in any other part of the nation.
    c. attitude change was solely responsible for this decline.
    d. all of the above
    e. none of the above

20. When racial equality has been achieved:
    a. all members of each group must be of the same status.
    b. the distribution of each group's members in the social structure is the same.
    c. a person's race does not reliably indicate his or her status.
    d. b and c
    e. a and b

Essay

1. A. Name three ways in which intergroup conflict may be resolved. (knowledge)
    B. Explain Allport's contact theory. (comprehension)
    C. Apply Allport's theory to a fictitious situation in a school or work setting. (application)

2. A. Name three ways in which minorities may achieve upward mobility. (knowledge)
    B. Explain economic enclave theory. (comprehension)
    C. Do the experiences of African-Americans seem to support or negate this theory? Explain. (application)

3. Explain the following statement: "Hispanic-Americans do not constitute an ethnic group; persons of Mexican, Puerto-Rican, and Cuban descent may each be considered distinct ethnic groups."

4. Trace the experiences of Japanese-Americans from their initial arrival in the United States to their position in American society today. Be certain to integrate into your explanation the various mechanisms by which minorities may achieve upward mobility.

5.   Trace the experiences of African-Americans in the United States from slavery to the present. Show how their position may be likened to that of recent immigrants.

6.   Recent research has shown that prejudice seems to be declining. These trends can be observed in movies and television shows. Today more films feature minority actors in major, less stereotypical roles than were evident a few decades ago. You may wish to observe the extent of these changes by watching films from the 1940's, 50's or 60's and comparing them with more recent ones. (Some of the stereotypes you see in the older films may shock or offend you but they were considered "normal" by many viewers in that time period.) Do you notice major changes and if so, what are some examples? Do you continue to see any evidence of stereotyping in current films? How might these changes both reflect and reinforce the decline of prejudice in our society?

Answers

Completion

1.	race	11.	democratic ideals, racist practices
2.	cultural heritages	12.	master, slave
3.	assimilation	13.	economic competitors
4.	accommodation	14.	a lack of information, economic motives
5.	extermination, expulsion, the imposition of a caste	15.	middleman
6.	prejudice	16.	recent immigrants
7.	authoritarian	17.	English
8.	Gordon Allport	18.	visibility
9.	intensify	19.	attitude change
10.	educated, income	20.	no homeland, numbers

Multiple-Choice

1.	e	11.	d
2.	e	12.	d
3.	d	13.	e
4.	a	14.	d
5.	d	15.	d
6.	a	16.	d
7.	c	17.	d
8.	a	18.	d
9.	b	19.	a
10.	d	20.	d

CHAPTER 12

GENDER AND INEQUALITY

Overview

Chapter 12 focuses on an analysis of gender and inequality using Guttentag and Secord's theory of sex ratios and sex roles. Many of the concepts introduced in earlier chapters are reviewed and applied to this analysis. Data from both Guttentag and Secord's study and from the author's analysis of data on gender inequality is provided throughout.

      The chapter opens with a brief description of gender inequality world-wide and then turns its attention to the Guttentag/Secord theory. It then applies the theory to the historical cases of Athens and Sparta. The concepts of dyadic power and structural power are then discussed and integrated into a more complete sociological explanation. The chapter then turns its attention to trends in North American sex ratios and moves to a discussion of the Woman Movement and the rise of feminism. It then examines women in the labor force and changing sexual norms. The chapter closes with an analysis of the impact of imbalanced sex ratios on the African-American family and the Hispanic-American family. Throughout the chapter, the reader gets a true "closer view" of theory construction and research in action.

Capsule Summary

Sex ratios, the number of males per 100 females, both affect and are affected by characteristics of the social structure. Guttentag and Secord argue that sex roles and relationships are strongly influenced by the ratio of males to females. Causes of imbalanced sex ratios include geographic mobility, female infanticide, health and diet, differential life expectancy, war, and sexual practices. In those societies where there is an excess of males, female status tends to be low, women are less likely to be employed outside the home, the divorce rate is low, and the birth rate is high. Historically, in such societies (for example, ancient Athens), women are considered the property of men and strict sexual norms restrict women's sexual conduct. Conversely, in societies where females outnumber males (our own, for example), sexual norms are relaxed, women are often employed outside the home, and women often unite to form social movements.

      Dyadic power refers to the capacity of each member of a dyad to impose his or her will on the other. An unfavorable sex ratio causes the member of the sex in excess supply to be in a position of power dependency. Structural power also plays a role in determining sex ratios and sex roles. In cases where males have lacked dyadic power, they have strengthened their structural power by imposing norms that severely restrict the freedom of women. In cases where an excess of females has caused males to have dyadic power, men have combined this with structural power to exploit women to the fullest.

      Guttentag and Secord argue that when one gender faces a lack of dyadic power, it will organize to remedy its problem. The Woman Movement and the rise of feminism in the nineteenth and twentieth centuries are examples of this.

Today, because women outnumber men, they increasingly work outside the home. Two-earner households are common, and occupations are less gender-segregated. Still, there is a gender gap in many fields. Men continue to earn higher wages, and many occupations (especially the high-paying ones) are still dominated by men. Job turnover rates are higher for women and sex-role socialization continues to play a role in limiting women's options.

The imbalanced sex ratio strongly affects the African-American family of today. Contrary to popular opinion, the African-American family was quite stable until very recently. Today, however, African-American men are in short supply. The illegitimacy ratio has increased, African-American males seem less willing to make commitments on the basis of sex, and the number of single-parent families headed by women has risen markedly.

The sex ratio among Hispanic-Americans differs somewhat according to country of origin. Factors such as age, differential immigration rates, and infant mortality rates partly explain these differences. Still, Hispanic-Americans continue to support the relationship between sex ratios and sex roles as postulated by Guttentag and Secord.

Key Concepts

You should be able to explain the concepts listed here and be able to cite several examples of each.

Sex ratio 352
Female infanticide 359
Dyadic power 363
Structural power 363
Feminism 372
Illegitimacy ratio 379

Key Theory

You should be able to explain the assumptions of this theory and be able to cite related research findings.

Guttentag and Secord: sex ratios and sex roles, 352
You should also be familiar with the data the author provides on gender inequality.

Completion

1. The number of males per 100 females is termed the _____.

2. In nations where men outnumber women, female status tends to be _____.

3. Where men outnumber women, the divorce rate is _____ and the birth rate is _____.

4. The major cause of imbalanced sex ratios is _____.

5. Where women are in excess, there is much _____ gender inequality.

6. _____ have higher fetal deaths and higher infant- and childhood-mortality rates.

7. Imbalanced sex ratios may be caused by geographic mobility, female infanticide, health and diet, war, _____, and _____.

8. Ancient Sparta is an example of a society where _____ outnumbered _____.

9. _____ refers to the capacity of each member of a dyad to impose his or her will on the other member.

10. Power based on statuses within social structures is termed _____.

11. The individual member whose sex is in short supply has a(n) _____ position and is _____ dependent on the partner because of the larger number of alternative relationships available.

12. When both _____ and dyadic power favor men, they respond by exploiting their advantage to the fullest.

13. In most societies, women tend to marry men somewhat _____ than themselves.

14. When one gender group faces a substantial lack of _____ for an appreciable time, it will organize to seek ways to remedy its problem.

15. When men are in _____ supply, women increasingly find the need to become self-supporting.

16. Although the majority of women now work outside the home, they bring home significantly _____ money than men do.

17. Because female employment rates have risen rapidly, the average working woman is considerably _____ than the average working man.

18. The percentage of all births that occur out of wedlock is termed the _____.

19. African-Americans have a substantially _____ infant-mortality rate than do whites.

20. Factors such as differences in age, infant-mortality rates, and _____ partly explain the differences in sex ratios among Hispanic-Americans from different countries of origin.

Multiple-Choice

1.   In nations where men outnumber women:
   a.   women are less likely to be employed outside the home.
   b.   the divorce rate is high.
   c.   the birth rate is low.
   d.   b and c
   e.   all of the above

2.   Imbalanced sex ratios may be caused by:
   a.   geographic mobility.
   b.   differential life expectancy.
   c.   sexual practices.
   d.   a and b
   e.   all of the above

3.   Female infanticide:
   a.   has rarely been practiced in non-Western societies.
   b.   is the major cause of imbalanced sex ratios.
   c.   has been unknown in Europe since 1500.
   d.   b and c
   e.   none of the above

4.   Compared with females, males:
   a.   have higher infant-mortality rates.
   b.   have a lower rate of fetal deaths.
   c.   have lower childhood-mortality rates.
   d.   b and c
   e.   none of the above

5.   Today, in most societies:
   a.   men outlive women.
   b.   women outlive men.
   c.   men and women have about the same life expectancy.
   d.   most men over the age of 65 are widowed.
   e.   a and d

6. Athenian sexual practices included:
    a. "respectable" women being isolated by extremely protective sexual norms.
    b. the highly visible presence of "disreputable" women.
    c. male homosexuality.
    d. all of the above
    e. none of the above

7. Among the characteristics of Spartan society were that:
    a. girls were offered as much education as boys.
    b. female infanticide was practiced but male infanticide was unknown.
    c. a Spartan wife was considered her husband's property.
    d. b and c
    e. none of the above

8. The capacity of each member of a dyad to impose his or her will on the other member is termed:
    a. structural power.
    b. dyadic power.
    c. dyadic advantage.
    d. structural advantage.
    e. none of the above

9. Dependent members of dyadic relationships may seek to improve their bargaining positions by:
    a. developing cultural means to make themselves unusually attractive to the opposite sex.
    b. spending more time in same-sex dyads.
    c. withdrawing from dyadic relationships.
    d. a and c
    e. all of the above

10. Men respond by exploiting their advantages to the fullest when:
    a. both structural and dyadic power favor women.
    b. structural power favors men but dyadic power favors women.
    c. both structural and dyadic power favor men.
    d. dyadic power favors men but structural power favors women.
    e. none of the above

11. The Woman Movement in the nineteenth and early twentieth centuries began and achieved its greatest prominence in _____ states.
    a. Midwestern
    b. Northeastern
    c. Southern
    d. Western
    e. none of the above

12. Features of the feminist ideology include:
    a. opposition to all forms of stratification based on gender.
    b. the belief that biology consigns females to inferior status.
    c. a sense of common purpose among women to direct their efforts to bring about change.
    d. a and c
    e. all of the above

13. Feminism:
    a. declined after the 1920s and was renewed in the 1960s.
    b. began in the 1960s.
    c. sprang from the Woman Movement and the suffrage movement in the second decade of the twentieth century.
    d. a and c
    e. all of the above

14. Due to the increased participation of women in the labor force:
    a. occupations have become less gender-segregated than they were in the past.
    b. gender segregation of occupations has disappeared.
    c. women bring home significantly less money than do men.
    d. a and c
    e. all of the above

15. As a result of the gender wage gap:
    a. women change jobs less often than men.
    b. the average working woman is considerably older than the average working man.
    c. women take time out from the labor market more often than men.
    d. all of the above
    e. none of the above

16. Sex-role socialization affects the gender gap because:
    a. women are more likely to be selected as leaders.
    b. women are less likely to enroll in majors that lead to high-paying jobs.
    c. women probably suffer from their smaller stature.
    d. b and c
    e. all of the above

17. The percentage of all births that occur out of wedlock is termed the:
    a. illegitimacy ratio.
    b. legitimacy ratio.
    c. fecundity ratio.
    d. crude birth rate.
    e. crude birth ratio.

18. The shortage of African-American males results from:
    a. higher infant mortality.
    b. the birth of more African-American females than African-American males.
    c. the high mortality of young African-American males from accidents, drugs, and violence.
    d. a and c
    e. all of the above

19. Which of the following is/are true about the African-American family?
    a. The African-American family was broken up by slavery.
    b. Sexual promiscuity was the norm on Southern plantations.
    c. The signs of severe disruption of the African-American family are very recent.
    d. a and b
    e. none of the above

20. Research has shown that:
    a. there are few, if any, differences in sex ratios for Hispanic-Americans from different countries of origin.
    b. Hispanic-American women, overall, face an extreme shortage of men.
    c. the role of women among Mexican-Americans is far less traditional than that among Puerto Rican-Americans.
    d. Hispanic-Americans do not support Guttentag and Secord's theory.
    e. none of the above

Essay

1. A. Explain dyadic and structural power. (comprehension)
    B. Show how these may be used to explain the outcome of imbalanced sex ratios. (application)
    C. Integrate these concepts and show how they would explain changing sex roles today. (analysis)

2. A. Name three causes of imbalanced sex ratios. (knowledge)
    B. Using the example of either Athens or Sparta, show how these sex ratios affected the role of women. (application)
    C. Contrast the roles of women in ancient Athens and Sparta using Guttentag and Secord's theory. (analysis)

3. Trace the Woman Movement and the rise of feminism from its beginnings until today.

4. Explain three reasons for the gender gap as it occurs in the work force today.

5. Discuss how Guttentag and Secord's theory explains the position of *either* the African-American family *or* the Hispanic-American family today.

6. Assume you attend a college where men outnumber women. You learn that your school is considering merging with a nearby college where women outnumber men. Using the Guttentag/Secord theory, how will this merger affect both the men and women who currently attend your college? What effects will it have on the students from the nearby school? If given a choice, would you support the merger? Do you think you would take the same position if you were of the opposite sex?

Answers

Completion
1. sex ratio
2. low
3. low, high
4. female infanticide
5. less
6. Males
7. differential life expectancy, sexual practices
8. women, men
9. Dyadic power
10. structural power
11. stronger, less
12. structural power
13. older
14. dyadic power
15. short
16. less
17. younger
18. illegitimacy ratio
19. higher
20. differential immigration rates

Multiple-Choice
1. a
2. e
3. b
4. a
5. b
6. d
7. a
8. b
9. e
10. c
11. b
12. d
13. c
14. d
15. c
16. d
17. a
18. d
19. c
20. e

CHAPTERS 9 TO 12

REVIEW AND SPECIAL PROJECT

Review

Chapters 9 through 12 have focused on social inequality and social stratification. They have discussed various sociological theories of stratification, different stratification systems, and the consequences of social inequality.

Special Project

The various social classes (and even groups occupying the different strata within the classes) often exhibit marked differences in life-style. Although these differences are not nearly as great as those that separated the classes in agrarian societies, they nevertheless continue to exist. Research has shown that the classes differ not only in obvious ways such as income and educational level but also in more subtle areas such as speech patterns, mortality rates, types of entertainment, and even preferences for particular brands of beer.

You may wish to speculate about certain aspects of life-style that might exhibit class differences. Choose one or two areas and devise specific hypotheses about the nature of these relationships. Search the literature for studies that test your hypotheses. (Bibliographies in sociology texts would be a good place to begin your literature search. You will also want to consult *Sociological Abstracts* during your study.)

If possible, you may also wish to obtain your own data through observation, interviews, or more unobtrusive measures such as noting television commercials and magazine advertisements geared to a target population. Once you have supported your hypothesis, you might explain this relationship by using the theories you have studied.

CHAPTER 13

THE FAMILY

Overview

Chapter 13 opens with a discussion of the dominant themes of the universality and the decline of the family. It then offers definitions of the term *family* and differentiates between nuclear and extended families. It discusses the functions of the family with cross-cultural comparisons before turning its attention to a discussion of the traditional European family. The theme of modernization (a major concern in later chapters) is discussed with emphasis on the transformation of the family and the effects of modernization on kinship and divorce. One-parent families are then discussed, and Patterson's work on the relationship between parenting practices and childhood deviance is described. (Again, the importance of attachments is emphasized in this research.) The chapter then discusses remarriage and concludes with Special Topic 4, the older family.

Capsule Summary

Families are small clusters of males and females, adults and children. Membership is typically determined by common ancestry and sexual unions. Although its form may differ, the family appears to be universal. Family forms include both the nuclear and the extended family. Traditionally, it was assumed the family performed the functions of reproduction, sexual regulation, economic cooperation, and education, but today the focus is more on the functions of sexual gratification, economic support, and emotional support. The incest taboo exists in every culture, although its form differs among cultures.

It has been assumed that the family is experiencing decline, although recent research by Shorter found the traditional European family to be almost the exact opposite of the large, close, and loving extended family it was long thought to be. Families were smaller than originally thought because infant- and childhood-mortality rates were high, children left home early, and adults often died before becoming old. Privacy was nonexistent; families lived in one room, which was often shared with outsiders. Attachments between parents and children were weak, and neglect was common. Relationships between spouses were indifferent or hostile. Peer group bonds were strong and served as the primary emotional attachments. Modernization brought about major changes. Romantic love rather than economic ties became the basis for marriage; emotional bonds between children and parents also strengthened. The family of today is a result of these changes.

Research by Trent and South attempts to explain why divorce is higher in some nations than in others. Divorce rates seem to be higher when modernization occurs, when a large proportion of women are employed outside the home, and when women outnumber men. Because approximately two-thirds of all divorces occur among couples who have children, much attention has been focused on the one-parent family. One-parent families result not only from divorce but also from the death of a spouse and from the <u>increasing number of single women becoming mothers</u>. Most one-parent homes are headed by <u>women</u>, and problems with <u>money</u> and <u>time</u> often plague the single parent.

It has often been assumed that children from one-parent homes are more prone to delinquency, but research in this area has produced mixed results. <u>Patterson</u>, in his study of <u>deviant children</u>, found that deviance is a result of <u>poor parenting</u> rather than of family structure. <u>Parents</u> of <u>deviant children</u> often <u>lack close attachments</u> to them and often <u>deny</u> or <u>fail to punish their deviant acts</u>.

More than <u>80 percent</u> of Canadians and Americans who divorce remarry. Recent research has been conducted in the area of <u>conjugal careers</u>, and studies such as those by <u>Jacobs and Furstenberg</u> and <u>White and Booth</u> have focused on the <u>economic characteristics</u> of <u>second spouses</u> and the <u>effects of stepchildren in the home</u> on the <u>likelihood of divorce</u>. Contrary to popular opinion, research has shown that the <u>psychological effects</u> of the <u>empty nest syndrome</u> seem to be quite <u>positive</u>.

Key Concepts

You should be prepared to explain the concepts listed here and be able to provide several examples of each.

Theme of the universality of the family 389	Family 391
	Nuclear family 392
Theme of the decline of the family 389	Extended family 392
	Functions of the family 393
	Incest taboo 393

Key Research Studies

You should be familiar with both the methodology and the results of the following research studies. The data on the older family appear in Special Topic 4.

World Values Survey: family functions 393

Shorter: *The Making of the Modern Family*; study of the traditional European family 397

Trent and South: international comparisons in divorce 410

Research on living together 411

Patterson: study of the relationship between parenting and childhood deviance 413

Jacobs and Furstenberg: second husbands; conjugal careers 415

White and Booth: stepchildren and marital happiness 415

NORC data on the older family 419

Completion

1. Sociological writing on the family has been dominated by the themes of _____ and _____.

2. Traditionally, the four primary functions of the family included sexual relationships, reproduction, _____, and _____.

3. A formal commitment to maintain a long-term relationship involving specific rights and responsibilities is termed _____.

4. An adult couple and their children are a(n) _____ family.

5. Virtually every society contains a(n) _____, which prohibits sexual relations between certain family members.

6. Family members such as the elderly and children who cannot support themselves are termed _____.

7. *The Making of the Modern Family* was written by _____.

8. The primary unit of sociability and attachment in the traditional European family was the _____.

9. The high divorce rate indicates that the marital relationship has become much _____ important than it used to be.

10. Approximately _____ of divorces occur among couples who have children.

11. The amount of extramarital sex that occurs is far _____ than is depicted in the media.

12. Trent and South concluded that divorce rates are _____ in societies where a large proportion of women are employed outside the home.

13. The proportion of children born to unwed mothers is termed the _____.

14. More than 90 percent of one-parent families are headed by _____.

15. Research has found that _____ is a primary cause of deviant behavior among children.

16. A major problem faced by female-headed families is lack of _____.

17. A major problem experienced by parents with poor parenting skills is a lack of _____ to their children.

18. White and Edwards have found that most parents experienced significant _____ in marital happiness when the last child moved out.

19. If there are no stepchildren in the home, couples who remarry _____ have a higher rate of divorce than people marrying for the first time.

*20. People older than 65 are almost _____ as likely as younger people to be fairly satisfied with their present financial situation.

*Drawn from Special Topic 4.

Multiple-Choice

1. Murdock defined the family as a social group characterized by:
    a. common residence.
    b. adults of both sexes.
    c. reproduction.
    d. a and c
    e. all of the above

2. Today the functions of the family typically include:
    a. sexual gratification.
    b. economic support.
    c. emotional support.
    d. a and b
    e. all of the above

3. An extended family is:
    a. larger than a nuclear family.
    b. composed of more than one nuclear family.
    c. composed of an adult couple and their children.
    d. a and b
    e. none of the above

4.	The theme of the decline of the family:
	a.	has been supported by recent research by Shorter.
	b.	argues that the functions of the family are best met by the modern family.
	c.	is obvious when the contemporary family is compared with the traditional family.
	d.	a and c
	e.	none of the above

5.	Which of the following is/are true about sexual gratification as a function of the family?
	a.	All societies have norms governing sexual behavior.
	b.	All societies have very narrow limits on who may engage in sex.
	c.	Sexual norms often change very rapidly.
	d.	a and c
	e.	all of the above

6.	Which of the following is/are research findings of Shorter's study?
	a.	The extended family living in a single household was typical regardless of economic level.
	b.	The traditional household was much smaller than had been assumed.
	c.	Female-headed households were very rare when compared with today's households.
	d.	all of the above
	e.	none of the above

7.	Shorter found that the size of the traditional family was smaller than previously assumed because:
	a.	infant and childhood mortality was high.
	b.	children often left home early.
	c.	adults often died before reaching "old age."
	d.	a and c
	e.	all of the above

8.	Shorter found that relations between husbands and wives and parents and children were:
	a.	based on emotional rather than economic ties.
	b.	often characterized by indifference or even hostility.
	c.	stronger than relationships to peer group members.
	d.	a and c
	e.	none of the above

9. In the traditional European family, *strong* emotional attachments were primarily between:
    a. persons of the same sex and peers outside the family.
    b. mothers and young children.
    c. husbands and wives.
    d. elderly parents and adult children.
    e. brothers and sisters.

10. Effects of modernization on the family include:
    a. an increase in privacy for families.
    b. a decrease in the importance of emotional ties as the basis for marriage.
    c. an increase in the importance of peer group relations.
    d. a and c
    e. none of the above

11. According to Table 13-9 (Family Ties):
    a. few adult Americans frequently socialize with living parents.
    b. almost two-thirds of married Americans rate their marriage as "very happy."
    c. as the family has grown smaller, family ties have severely weakened.
    d. a and c
    e. none of the above

12. More than _____ of those who divorce remarry.
    a. 10 percent
    b. 50 percent
    c. 80 percent
    d. 90 percent
    e. 98 percent

13. In their study of international comparisons in divorce, Trent and South found that:
    a. as nations continue to develop, the divorce rate soon begins to decline.
    b. the divorce rate is higher where a large proportion of women work outside the home.
    c. the larger the proportion of Catholics, the lower the divorce rate.
    d. b and c
    e. all of the above

14. Research on living together has found that:
    a. women who lived with their eventual husbands were more likely to get divorced than women who did not.
    b. living together seemed to result in a reduced divorce rate.
    c. living together for more than three years before marriage greatly reduces the divorce rate.
    d. b and c
    e. none of the above

15. Patterson found that deviant children:
    a. had always been raised in one-parent families.
    b. were often greatly loved by their parents but that this attachment was not returned.
    c. often had parents with poor parenting skills.
    d. a and b
    e. none of the above

16. Patterson found that parents of problem children:
    a. have weak attachments to their children.
    b. often fail to punish their children for failing to obey.
    c. often refuse to "see" what their children are doing.
    d. all of the above
    e. none of the above

17. According to Jacobs and Furstenberg:
    a. women overwhelmingly better their economic situation by remarriage.
    b. women, on average, marry second husbands who are no more and no less successful than their first husbands.
    c. a woman with children younger than 10 is in the best position to improve her economic situation by remarriage.
    d. a and c
    e. a and b

18. According to research by White and Booth:
    a. stepchildren leave home at a younger age than do biological children.
    b. couples who remarry have a higher divorce rate than do people who are marrying for the first time if there are no stepchildren in the home.
    c. couples who remarry do not have a higher rate of divorce than do people who are marrying for the first time if there are no stepchildren in the home.
    d. a and b
    e. a and c

19. Research on the "empty nest syndrome" by Edwards and White found that:

a. many parents, especially mothers, experience serious emotional problems when their children leave home.
b. most parents experienced significant increases in marital happiness when the last child had moved out.
c. the resulting psychological effects tended to be quite positive.
d. b and c
e. none of the above

*20. Research on the older family has shown that:
a. older Americans are the happiest Americans.
b. older Americans are overwhelmingly dissatisfied with their income.
c. older Americans are less likely to attend church weekly than are younger Americans.
d. all of the above
e. none of the above

*Drawn from Special Topic 4.

Essay

1. A. Name the four functions of the family. (knowledge)
   B. Discuss the theme of the universality of the family. (comprehension)
   C. Contrast the nuclear family with the extended family. (analysis)

2. A. Discuss some of the characteristics of the traditional European family as described by Shorter. (comprehension)
   B. Contrast the typical view of the extended family with the traditional family described by Shorter. (application)
   C. Relate the theme of decline to your discussion. (analysis)

3. Explain the following statement: "The current high divorce rate . . . could mean that at any given moment the great majority of marriages are happy ones."

4. Describe the changes in the family brought about by modernization.

5. Discuss five research findings about older Americans as described in Special Topic 4.

6. Assume you have taken a part-time (or summer) job working with young children. You notice that some children are better behaved than others. How can you use your knowledge of the effects of single parent homes, stepfamilies, and Patterson's work on parenting techniques to partially explain these differences? Could you devise some techniques to work with these children on the basis of your information? If you were asked to assist with a "parent involvement" program what areas would you suggest be included to help these parents better relate to their children?

Answers

Completion
1. universality, decline
2. economic cooperation, education (socialization)
3. marriage
4. nuclear
5. incest taboo
6. dependents
7. Edward Shorter
8. peer group
9. more
10. two-thirds
11. less (or lower)
12. higher
13. illegitimacy ratio
14. females
15. poor parenting
16. income
17. attachment
18. increases
19. do not
20. twice

Multiple-Choice
1. e
2. e
3. d
4. e
5. d
6. b
7. e
8. b
9. a
10. a
11. b
12. c
13. b
14. a
15. c
16. d
17. b
18. e
19. d
20. a

CHAPTER 14

RELIGION

Overview

Chapter 14 opens with a discussion of the nature of religion and how it differs from other systems of explanation. It describes how religion makes norms legitimate and serves as a major force in creating moral communities, introduces the concept of a religious economy, and considers church-sect theory in depth. The chapter discusses secularization and its effects on both sect revival and cult formation. It also discusses the American and Canadian religious economies and cites numerous research findings by the author and others about sect and cult formation and concentration in both the United States and Europe. The Protestant explosion in Latin America is examined. After discussing the universal appeal of religion, the chapter concludes with a discussion of the revival of religion in Russia following the break-up of the former Soviet Union. The importance of attachments is a recurrent theme in this chapter, and considerable attention is paid to the relationship between attachments and religious application.

Capsule Summary

Religion consists of socially organized patterns of beliefs and practices concerning questions of ultimate meaning that assume the existence of the supernatural. Evidence that religion existed 100,000 years ago has been found in the remains of Neanderthal culture. Religion serves to make norms legitimate by explaining why they exist and should be followed.

As societies grew more complex, many different religions emerged, often side by side, thus creating pluralistic religious economies. Weber distinguished between the concepts of church and sect. Further elaborations of church-sect theory by Niebuhr and Johnson postulate that churches are characterized by intellectualized teachings, emotional restraint in services, and deities who are remote from human affairs. They exist in a state of relatively low tension with the sociocultural environment. Sects, on the other hand, embrace emotionalism, fundamentalism, and personal relationships with deities. They tend to exist in a state of high tension with the sociocultural environment.

Secularization involves turning away from religion to a secular view of life. Initially, many people thought that increased secularization would lead to the eventual decline of religion. Recent research has not supported this argument. Although secularization does lead to a decline in dominant religious organizations such as churches, sects and cults often emerge to take their place. Sects differ from cults in that sects are revivals of an old religion. (The term revival is used to indicate sect formation.) Cults, on the other hand, are new religions and emerge as a result of innovation. Charisma is the ability of some people to inspire faith in others, and charismatic leaders are often very instrumental in religious movements.

The North American religious economy is very diverse. Church attendance is high, and almost two-thirds of all North Americans are members of a local congregation. Sect formation is common because many of the dominant organizations are experiencing decline. Sect movements are clustered where church membership is highest because they represent a revival of an already established religion. Because cults represent a new religion, they experience greatest success in areas where church attendance is low. Cults often attract people who have no previous religious affiliation. Thus, rather than leading to a decline in religion itself, secularization has led to sect and cult formation as a response to the decline of the dominant religious organizations. Similar findings have been reported in Europe, thus further strengthening this cyclical theory. Recently, sociologists have noticed a major increase in Protestantism in Latin America as it too becomes a competitive religious economy. Likewise there has been a major revival in religion in Russia following the break-up of the former Soviet Union.

Key Concepts

You should be able to explain the concepts listed here and be prepared to cite several examples of each.

Religious economy 427	Church 433
Revival 428	Sect 433
Questions of ultimate meaning 429	Innovation (cult formation) 433
Religion 429	Secularization 435
Supernatural 429	Cult 439
Religious pluralism 433	Charisma 440

Key Research Studies

You should be familiar with both the methodology and the results of these research studies.

World Values Survey: religiousness around the world 431

Stark, Bainbridge, and others: relationship between church membership and various forms of deviation; application of church-sect theory to the concept of a religious economy; studies of rates of church membership, religious affiliation, and cult formation 436

Melton et al.: cult movements in Europe 448

Data on the increase of Protestantism in Latin America 451

Greeley (and others): religious revival in the former Soviet Union 452

You should be able to explain the assumptions of these two theories and, when applicable, cite related research findings.

      Church-sect theory (Niebuhr and Johnson)

      Religious economy

_____

Completion

1. The marketplace of competing faiths within a society is termed a(n) _____.

2. All religions involve answers to questions about _____.

3. Religion entails socially organized patterns of beliefs and practices concerning ultimate meaning and assumes the existence of the _____.

4. The natural state of a(n) _____ is pluralism.

5. _____ first distinguished between churches and sects.

6. _____ intellectualize religious teachings and restrain emotionalism in their services.

7. Sects stress _____ and individual mystic experience and tend toward _____.

8. Niebuhr argued that _____ provide for the religious needs of persons low in the stratification system.

9. Johnson suggested that church and sect are opposite poles on an axis representing the degree of tension between religious organizations and their _____.

10. Churches are religious bodies with relatively _____ tension, whereas sects are religious bodies with relatively _____ tension.

11. A turning away from religious to secular explanations of life is termed _____.

12. The process of the formation of sects is termed _____.

13. Revival is often a response to _____.

14. A religious movement that represents a new and unconventional faith is termed a(n) _____.

15. Weber termed the ability of some people to inspire faith in others _____.

16. Membership in conventional religious groups will be highest where _____ are most active.

17. Research by Melton, Stark, and others has found that European nations have twice as _____ a rate of cult movements as the United States.

18. As religious bodies deemphasize the _____, they seem less able to satisfy religious needs.

19. Studies have found that _____ abound in places where church attendance is lowest.

20. _____ do best when they tap into a strong religious tradition, whereas _____ abound where the conventional religious tradition is weak.

Multiple-Choice

1.   Most sociologists agree that religion always entails:
       a.   answers to questions of ultimate meaning.
       b.   the existence of the supernatural.
       c.   the belief in one supreme deity.
       d.   a and b
       e.   all of the above

2.   Religious institutions:
       a.   can be a major force in holding societies together.
       b.   can give legitimacy and reason to the norms.
       c.   can give divine sanctions to other institutions.
       d.   a and b
       e.   all of the above

3.   The natural state of a religious economy is:
       a.   pluralism.
       b.   oligarchy.
       c.   monopoly.
       d.   monarchy.
       e.   anarchy.

4.   Churches:
       a.   intellectualize religious teachings.
       b.   stress emotionalism.
       c.   tend toward fundamentalism in their teachings.
       d.   b and c
       e.   all of the above

136

5. Sects:
    a. represent the gods as close at hand.
    b. stress individual mystical experiences.
    c. restrain emotionalism in their services.
    d. a and b
    e. all of the above

6. Niebuhr argued that sects provide for the religious needs of people:
    a. of low status.
    b. in the middle class.
    c. in the upper class.
    d. with high educational levels.
    e. b and c

7. Johnson argued that _____ are religious bodies with relatively high tension.
    a. cults
    b. sects
    c. churches
    d. a and b
    e. all of the above

8. Traditionally, many social scientists believed that secularization would:
    a. lead to the eventual disappearance of religion.
    b. increase membership in churches rather than in sects.
    c. foster the rise of cults.
    d. increase membership in sects.
    e. c and d

9. The formation of sects is termed:
    a. revelation.
    b. revival.
    c. secularization.
    d. pluralism.
    e. innovation.

10. Responses to secularization include:
    a. revivals.
    b. cults.
    c. religious change.
    d. a and b
    e. all of the above

11. Sects are:
    a. new religions.
    b. based on religions outside the conventional religious tradition.
    c. new organizations reviving an old religion.
    d. a and b
    e. none of the above

12. For *contemporary* sociologists of religion, the basis of charisma is:
    a. the ability to inspire faith in others.
    b. the ability to get others to believe one's message.
    c. the unusual ability to form attachments with others.
    d. a and b
    e. none of the above

13. Research has found that almost _____ of North Americans are official members of a local congregation.
    a. four-fifths
    b. three-quarters
    c. two-thirds
    d. one-half
    e. one-third

14. Sect movements are clustered in those states where:
    a. membership in Christian churches is low.
    b. membership in Christian churches is high.
    c. church attendance is low.
    d. a and c
    e. none of the above

15. Research by Melton, Stark, and others has found that:
    a. previous research has significantly undercounted European religious groups.
    b. European religious groups have previously been studied more often than religious groups in the United States and Canada.
    c. cult movements are far rarer in Europe than in the United States.
    d. a and c
    e. none of the above

16. In the United States, religious innovation is more common and successful in _____ than elsewhere in the nation.
    a. the Deep South
    b. the Far West
    c. the Northeast
    d. the Middle West
    e. New England

17. In contrast to the United States, the Canadian religious economy:
    a. is characterized by low church membership.
    b. is more diverse.
    c. is less diverse.
    d. has experienced a decline in sectlike groups and a growth in denominations.
    e. none of the above

18. Nock's Canadian data found that:
    a. all Canadian sects do best where religious affiliation is highest.
    b. conventional Protestant sects were strongest in provinces where the people were least likely to say they had no religion.
    c. conventional Protestant sects were strongest in provinces where cult movements were doing best.
    d. all Canadian sects do best in a more secularized environment.
    e. none of the above

19. Research on religion in Latin America has shown that:
    a. there is an increase in Protestants, especially Pentecostal groups.
    b. there is an increase in Protestants, especially denominational groups.
    c. there is a decline in the comparative religious economy as more and more people embrace Catholicism.
    d. fewer people of any faith are active in religion.
    e. none of the above

20. According to Greeley's data on the revival of religion in Russia:
    a. the average citizen of the Russian federation is the "scientific atheist" the schools have created.
    b. the new converts are concentrated among the younger and better educated Russians.
    c. the new converts are concentrated among the older and less educated Russians.
    d. a and c above
    e. none of the above

Essay

1. A.   Define religion. (knowledge)
   B.   Discuss the nature of religion. (comprehension)
   C.   Contrast religion with other systems of explanation. (analysis)

2. A.   Define a religious economy. (knowledge)
   B.   Explain the following statement: "The natural state of a religious economy is pluralism." (comprehension)
   C.   Apply the concept of a religious economy to both the United States and Canada. (application)

3. Explain Niebuhr's church-sect theory. Discuss the later modifications of this theory.

4. How does the process of secularization lead to sect and cult formation?

5. Distinguish between a sect and a cult. Discuss two research findings about geographical concentrations of sects and cults.

6. Many people have a very negative image of cults and cult members. This image has been partly fostered by the media in their coverage of rare but "sensational" tragedies such as those in Waco, Texas in 1993 or the mass suicides and murders in Jonestown in 1978.

   Prior to reading this chapter did you share this image? Have your views changed as a result of this new information? (If you are a member of a cult, have you ever experienced prejudice and/or discrimination as a result of your membership?) How could you use the information you obtained from this chapter to help educate the general public and reduce some of these negative stereotypes?

Answers

Completion

1.	religious economy	11.	secularization
2.	ultimate meaning	12.	revival
3.	supernatural	13.	secularization
4.	religious economy	14.	cult
5.	Max Weber	15.	charisma
6.	Churches	16.	sects
7.	emotionalism, fundamentalism	17.	high
8.	sects	18.	supernatural
9.	sociocultural environment	19.	cults
10.	low, high	20.	Sects, cults

Multiple-Choice

1.	d	11.	c
2.	e	12.	c
3.	a	13.	c
4.	a	14.	b
5.	d	15.	a
6.	a	16.	b
7.	b	17.	e
8.	a	18.	b
9.	b	19.	a
10.	e	20.	b

CHAPTER 15

POLITICS AND THE STATE

Overview

Chapter 15 begins with a discussion of the medieval practice of the "freedom of the commons" and traces how population changes led to the "tragedy of the commons." It then discusses the concept of collective goods and describes the functions of the state. It traces the rise of the state and describes efforts in both England and the United States to "tame the state." It examines elitist and pluralist states and focuses on Mills's concept of the power elite. The chapter then turns its attention to public opinion and focuses on the Gallup Poll. Research on gender and candidate success is discussed, and the chapter closes with a consideration of ideology and its role in political behavior.

Capsule Summary

States function to preserve internal order, maintain external security, and provide for collective goods. Although some people have argued in favor of anarchy, most feel that the state is a necessary aspect of complex society. Because states use coercion, "taming the state"—that is, limiting its powers—has been a dominant historical issue. The history of England and the United States contains numerous examples of this process, which entails the establishment of both a clear set of rules defining the limits of power and a structure designed to ensure that the rules are observed.

      States are of two essential types: elitist and pluralist. Elitist states, which are characterized by the rule of a single elite (sometimes termed the power elite), are the most common type. Pluralist states, on the other hand, are composed of many elites competing for power. In such states, power is distributed among various shifting coalitions.

      Public opinion polls, notably the Gallup Poll, are key indicators of public opinion about political issues. They are used extensively today to tap information about voter preference. Recent research has focused on the relationship between gender and voter preference. Although many have argued that female candidates may not be successful because of their gender, research by Hunter and Denton and Newman has found that the relationship between gender and vote getting is spurious and other factors are influential in this pattern.

      Ideologies, or theories about how societies should be run, do not enjoy the popular appeal in the United States and Canada that they do in Europe. Indeed, the results of panel studies have shown that Americans do not often maintain the same ideological conviction over time. Political interest here often takes the form of interest in a specific issue rather than in an underlying ideology, and the term issue public has been used to identify groups actively involved in a particular issue. Indeed, rather than being strongly ideological like their European counterparts, successful political parties in the United States actually are coalitions of many internal issue publics.

Key Concepts

You should be prepared to explain these concepts and be able to cite several examples of each.

"Freedom of the commons" 457
Public goods (collective goods) 459
State 463
Anarchy 464
Pluralism 467
Tyranny of the minority 468
Tyranny of the majority 468

System of checks and balances 468
Elitist state 469
Pluralist state 469
Power elite 471
Ideology 481
Panel study 485
Issue publics 485

Key Research Studies

You should be familiar with both the methodology and the results of these research studies.

Messick and Wilke: laboratory recreation of "freedom of the commons" 459
World Values Survey: interest in politics 472
Gallup Poll data on political participation 474
Hunter and Denton: gender and vote-getting ability 479
Newman: comparison of the success of men and women candidates 480
McClosky: elites and ideology 484

Key Theories

You should know how to explain the assumptions of these two theories and, when applicable, cite related research findings.

Pluralist theory
Development of the state (not literally a theory)

Completion

1. Olsen argued that to create _____ or _____ goods, the interests of the individual and the interests of the group collide.

2. The _____ or _____ is the organized embodiment of political processes within a society.

3. Only through organized _____ can humans ensure themselves of public goods.

4. The practice in medieval England of allowing tenants to use all uncultivated pasture lands was termed _____.

5. In his book *The Leviathan*, Thomas Hobbes describes what life would be like in a condition of _____.

6. For people to live in groups, internal order must be maintained, protection must be secured from external dangers, and _____ must be provided.

7. The existence of states rests on the development of _____.

8. _____ occurs when political power is dispersed among groups with diverse interests.

9. The danger that a majority of citizens will use the machinery of representative government to exploit and abuse minorities is termed _____.

10. The system known as _____ ensures that within the three branches of government each branch has the power to nullify actions taken by the other two.

11. The _____ state is the most common type.

12. In a(n) _____ state, power is maintained by the existence of many competing elites.

13. Mills argued that the United States is effectively ruled by a small set of influential people termed the _____, who hold the preponderance of power.

14. Research by Hunter and Denton has found that the relationship between gender and vote getting is _____.

15. Newman's research has found a candidate's sex _____ his or her chances of winning an election.

16. A connected set of strongly held beliefs based on a very few abstract ideas is termed a(n) _____.

17. Converse used the term _____ to identify those who take interest in and who participate at least as observers in discussions of an issue.

18. Successful parties are coalitions of many internal _____.

19. Compared with politics in Europe, politics in the United States is _____ very ideological.

20. The first successful effort to conduct a public opinion poll that correctly predicted the outcome of a presidential election was led by _____.

Multiple-Choice

1.   For a group to survive:
      a.   internal order must be maintained.
      b.   it must be secure from external dangers.
      c.   public goods must be provided.
      d.   a and b
      e.   all of the above

2.   Research by Messick and Wilke ("freedom of the commons" simulation) found that:
      a.   subjects tended to use their power to exploit others.
      b.   subjects behaved markedly different than had the English lords.
      c.   leaders in the experiment gave themselves smaller shares than they gave others.
      d.   all of the above
      e.   none of the above

3.   Which of the following statements is/are true about the rise of the state?
      a.   The existence of states rests on the development of agriculture.
      b.   Agrarian states have a low degree of stratification.
      c.   In small, simple societies, the state is often a loosely organized authority structure based on kinship and age.
      d.   a and c
      e.   all of the above

4.   As societies become more complex:
      a.   the machinery of state becomes less elaborate but more specialized.
      b.   the machinery of state becomes less elaborate and less specialized.
      c.   the machinery of state becomes more elaborate and more specialized.
      d.   fewer people hold positions as full-time leaders.
      e.   none of the above

5. In a condition of _____, political power is dispersed among groups with diverse interests.
    a. monarchy
    b. pluralism
    c. anarchy
    d. authoritarian rule
    e. communalism

6. _____ theory of the state holds that private property is the root of all repression and exploitation by the ruling class.
    a. Marxist
    b. Pluralist
    c. Anarchist
    d. Functionalist
    e. Weberian

7. Madison termed _____ the danger that a privileged few would use the machinery of representative government to exploit and abuse the many.
    a. tyranny of the majority
    b. tyranny of the minority
    c. anarchy
    d. pluralist tyranny
    e. none of the above

8. In a pluralist state:
    a. rules governing state power are maintained by the existence of many competing elites.
    b. all persons living in the state have an equal amount of power in decision making.
    c. shifting coalitions of many minorities rule.
    d. a and c
    e. none of the above

9. According to Mills, the power elite in the United States is predominantly:
    a. Protestant.
    b. male.
    c. educated in Ivy League schools.
    d. a and c
    e. all of the above

10.   The rise of opinion polling is most closely associated with:
       a.   Thomas Hobbes.
       b.   George Gallup.
       c.   C. Wright Mills.
       d.   Thomas Jefferson.
       e.   David Reisman.

11.   According to research by Hunter and Denton (gender and vote getting):
       a.   political parties start losing because they nominate women.
       b.   after political parties start losing they increase their rate of female
            nominations.
       c.   the effect of gender on vote getting may be spurious.
       d.   b and c
       e.   all of the above

12.   Which of the following is/are possible reasons why women may not be fully
       represented among office holders?
       a.   Journalists, party leaders, and potential candidates believe voters prefer
            male candidates.
       b.   Political action committees may be reluctant to fund female candidates.
       c.   Too few women run for office.
       d.   a and b above
       e.   all of the above

13.   An ideology is:
       a.   a connected set of strongly held beliefs based on a few abstract ideas.
       b.   a guide to one's reaction to external events.
       c.   essentially a theory about life.
       d.   b and c
       e.   all of the above

14.   A study in which the same sample of respondents is interviewed several times
       is termed a(n) _____ study.
       a.   retrospective
       b.   case
       c.   panel
       d.   ex post facto
       e.   none of the above

15. Converse used the term _____ to identify those who take an interest in and who participate at least as observers in discussions of an issue.
    a. interest groups
    b. ideology supporters
    c. interest publics
    d. issue publics
    e. support groups

16. Many people lack political ideologies because:
    a. they rarely invent their own.
    b. ideologies are intellectual creations often involving many different authors and interpreters.
    c. only elites can create and preserve ideologies.
    d. all of the above
    e. none of the above

17. In the United States and Canada, successful political parties:
    a. are strongly ideological.
    b. appeal to a narrow interest group within an electorate.
    c. are coalitions of many internal issue publics.
    d. a and b
    e. none of the above

18. Newman's study of the relationship between sex and candidate success has found that:
    a. incumbents have a large advantage in elections.
    b. a candidate's sex strongly affects his or her chances of winning an election.
    c. current office holders are disproportionally men.
    d. a and c above
    e. all of the above

19. Elitist states:
    a. are nonexistent today.
    b. never call themselves democracies.
    c. are the most common type.
    d. b and c
    e. none of the above

20.   Who of the following is not correctly paired with his contribution?
   a.   Thomas Hobbes: *The Leviathan*
   b.   James Madison: system of checks and balances
   c.   George Gallup: the desirability of anarchy
   d.   Plato: concept of the philosopher-king
   e.   none of the above

Essay

1.   A.   Name the two types of states. (knowledge)
     B.   Give examples of elitist and pluralist states. (comprehension)
     C.   Contrast elitist states with pluralist states. (analysis)

2.   A.   Name three functions of the state. (knowledge)
     B.   Explain Olsen's argument about the necessity of coercion. (comprehension)
     C.   Apply his argument to the "taming of the state." (application)

3.   Trace the taming of the state in England and in the United States.

4.   What is a political ideology? Compare and contrast political and religious ideologies. Explain the following statement: "Compared with Europe, politics in the United States is not very ideological."

5.   Explain the following statement: "Election day results do not mirror public opinion."

6.   Do you think the research on sex and candidate success also applies to major positions in student government on your campus? Have men (or women) predominated among major office holders in the past few years? If so, do you notice any recent changes? In what ways are campus politics similar to those at state and national levels? In what ways might you expect them to differ? If men have predominated, what advice would you give a freshman or sophomore woman who aspires to hold a major office in her junior or senior year?

Answers

Completion

1. public, collective
2. state, government
3. coercion
4. freedom of the commons
5. anarchy
6. public (collective) goods
7. agriculture
8. Pluralism
9. tyranny of the majority
10. checks and balances
11. elitist
12. pluralist
13. power elite
14. spurious
15. does not affect
16. ideology
17. issue publics
18. issue publics
19. not
20. George Gallup

Multiple-Choice

1. e
2. a
3. d
4. c
5. b
6. a
7. b
8. d
9. e
10. b
11. d
12. e
13. e
14. c
15. d
16. d
17. c
18. d
19. c
20. c

CHAPTER 16

THE INTERPLAY BETWEEN
EDUCATION AND OCCUPATION

Overview

After opening with a discussion of the interplay between education and occupation, Chapter 16 discusses the prestige rankings of occupations. It examines the changing nature of work and the composition of the labor force. It discusses unemployment and focuses on the relationship between the changing nature of work and rates of unemployment. It then considers the history of education in the United States and the current concern with the decline in the quality of education. The issue of the importance of schools and the related research of Coleman and Heyns are examined. The chapter closes with an in-depth discussion of Meyer's theory of educational functions.

Capsule Summary

The interplay of education and occupation is a dominant feature of societies. Education has always been highly valued in the United States. Its importance has increased as we have moved from a predominantly industrial economy to a knowledge economy. Typically, those whose educational level is high enjoy not only greater income but also greater prestige; a person's occupation is a major source of prestige. Prestige rankings, which tend to be relatively consistent over time and place, have shown that the more training or skill required for an occupation, the higher its prestige.

During this century, the nature of work and education has changed. More positions require skill and extensive training, and fewer rely on physical labor. Hence, unemployment rates for the unskilled are high. Women have entered the work force in greater numbers, partly as a response to the changing nature of work. More students are staying in school longer; high school graduation is commonplace, and the majority of Americans enter college. Nevertheless, Americans are concerned with the quality of education. Recent research has indicated that many Americans, including college graduates, do not score well on literacy tests. There is also evidence that students in public schools are more likely to drop out, do less homework, and have fewer plans to go to college than do students in Catholic and private schools.

Research has focused on the functions of schools. It was long assumed that the quality of a school would have an effect on learning. Coleman's study, however, found that school quality had no detectable impact on student achievement scores.

Heyns found that summer vacation seemed to be most detrimental to lower-income children, who probably benefit the most from school. Further research by Alexander et al. and Heyneman and Loxley has produced additional evidence of the importance of education for the disadvantaged. There has been recent concern with the "devaluation of education." Collins argues that the increasing emphasis on higher education may be creating a "credential" society. Meyer's theory of educational functions argues for the importance of education as a socializing agent. He concludes

151

that the <u>main function of education</u> is to <u>confer prestigious statuses</u> and to train persons to play the roles attached to them.

Similarly, colleges have the power to create new positions that are accorded high prestige. For the occupants of these statuses, then, the prestige and life-style associated with them does not end on graduation but rather becomes a lifelong identity.

Key Concepts

You should be ready to explain the concepts listed here and be able to give several examples of each.

Occupational prestige 492

Scientific management 496

Unemployment 499

"Credential society" 517

Key Research Studies

You should be familiar with both the methodology and the results of these research studies.

Hatt and North: occupational prestige rankings in the United States 492

Porter and Pineo: occupational prestige rankings in Canada 493

Taylor: time and motion studies 495

General Social Survey: job satisfaction 500

Kirsch and others: E.T.S. literacy studies 505

Coleman et al.: quality of schools and student achievement scores 508

Heyns: the effects of summer vacation on learning 508

Data from "High School and Beyond" study (prepared by Stark) 511

Heyneman and Loxley: school effects worldwide 514

Key Theories

You should know how to explain the assumptions of these two theories and, when applicable, cite related research findings.

Allocation theories

Meyer's theory of educational functions

The following are not exactly theories per se, but knowledge of these trends and changes is essential for understanding Chapter 16.

      The history of education in the United States

      Changes in the composition of the work force

      "Inflation" in higher education (Collins et al.)

Completion

1. Generally, the more education people have, the _____ they earn and the _____ their occupational status.

2. The more training an occupation requires and the more pay it offers, the _____ its public prestige.

3. The application of scientific techniques to improve work efficiency is termed _____.

4. Technological innovations have made it possible to work _____.

5. According to Drucker, we are changing from a primarily industrial economy to a(n) _____ economy.

6. The term _____ is applied to people 16 years of age and older who are without jobs and are seeking work.

7. Coleman found that school quality had little impact on student _____.

8. Collins argued that education was not meant to prepare people for careers but to protect _____.

9. Heyns found that schools greatly improve the situations of _____ children.

10. Heyns found that the single activity that is most strongly and consistently related to summer learning is _____.

11. Research by Alexander et al. found that dropping out of high school had the most severe negative effects on students from the most _____ backgrounds.

12. The _____ the nation, the greater the economic returns for getting an education.

13. As the level of education has risen in industrial nations, the relative advantage of completing a given level of education has _____.

14. Meyer argued that variations in school quality seem of _____ importance in the attitudes, values, opinions, and behaviors of graduates.

15. Meyer argued that the real impact of schools is to admit people to a particular
_____.

16. _____ theories argue that education is a passive servant of the stratification system.

17. Meyer argued that education helps to create new classes of _____ and _____, both of which are then incorporated into society.

18. As more people get an education, a given level of education becomes _____ valuable.

19. The respect given a person on the basis of his or her job is termed _____.

20. Those people who are employed or seeking employment make up the _____.

Multiple-Choice

1.　　Studies of occupational prestige:
　　a.　yield results that are fairly stable over time and place.
　　b.　have shown that many of the higher-prestige positions require a college education.
　　c.　have found that Canadians have a markedly different ranking system than do Americans.
　　d.　a and b
　　e.　b and c

2.　　The interplay between education and occupation:
　　a.　begins in adolescence.
　　b.　begins early in life.
　　c.　does not manifest itself until adulthood.
　　d.　is declining as a result of "working smarter."
　　e.　none of the above

3.　　Drucker has argued that:
　　a.　modern workers work harder and smarter than did their grandparents.
　　b.　we are changing from a primarily industrial economy to a knowledge economy.
　　c.　we are changing from a knowledge economy to an industrial economy.
　　d.　a and b
　　e.　none of the above

4.   Reasons for the increased participation of women in the work force include:
       a.   the feminist movement.
       b.   reduced fertility.
       c.   a change in the kinds of work available.
       d.   a and b
       e.   all of the above

5.   The labor force has expanded because:
       a.   a greater proportion of young people are working today.
       b.   a greater proportion of older people are working today.
       c.   women have entered the work force.
       d.   a and c
       e.   all of the above

6.   The term *unemployed* includes only:
       a.   people 16 and older.
       b.   people without jobs.
       c.   people who are seeking work.
       d.   a and b
       e.   all of the above

7.   Reasons for the high rates of unemployment among African-Americans include:
       a.   discrimination.
       b.   that African-Americans today are far less likely to enter college than are whites.
       c.   the dwindling supply of unskilled labor jobs.
       d.   a and c
       e.   all of the above

8.   Coleman's study found that:
       a.   school quality had a major impact on student achievement scores.
       b.   teachers' educational levels had a major impact on student achievement scores.
       c.   school quality did not have a major impact on student achievement scores.
       d.   summer vacation had little effect on students' achievement.
       e.   a and b

9. Today approximately _____ of Americans are in the labor force.
   a. 64 percent
   b. 75 percent
   c. 40 percent
   d. 88 percent
   e. 51 percent

10. Recent research on literacy has found that:
    a. approximately one half of all Americans scored in the lower two categories of each form of literacy.
    b. the low scores were concentrated among people who drop out of high school or are recent immigrants who have a limited ability to speak English.
    c. the vast majority of college graduates scored in the highest categories of quantitative, prose, and document literacy.
    d. a and c above
    e. all of the above

11. Heyns found that schools:
    a. merely maintain the differences that poor and middle-class children bring to schools.
    b. greatly improve the situations of poor children.
    c. greatly improve the situations of middle-class children but have little effect on poor children.
    d. have little effect on either poor or middle-class children.
    e. none of the above

12. Heyn's study of the effect of summer vacation found that:
    a. attending summer school prevents summer learning losses.
    b. children from all levels were harmed by summer vacations.
    c. children from higher-income families learned about as much during vacation as they did during the school year.
    d. a and b
    e. none of the above

13. Research by Heyneman and Loxley in school effects in twenty-nine nations found that:
    a. the poorer the nation, the less that students' backgrounds influence their school performances.
    b. children in less industrialized nations learn more during the same number of school years than do those in more industrialized nations.
    c. the poorer the nation, the less the economic returns for getting an education.
    d. a and b
    e. all of the above

14. The decline of the value of a college education can be attributed to:
    a. the result of colleges not preparing people for careers.
    b. the rising relative earnings of blue-collar workers, which have surpassed the earnings of some college graduates.
    c. the fact that college graduates are no longer a scarce commodity.
    d. b and c
    e. all of the above

15. Meyer argued that:
    a. a major effect of education is that people learn to play the role appropriate to the status their schools confer on them.
    b. the most powerful socializing property of schools is their ability to confer statuses that are recognized in the society at large.
    c. educational institutions have the power to create new occupations and to control the placement of these occupations in the occupational structure.
    d. all of the above
    e. none of the above

16. Allocation theorists argue that the primary purpose of education is to:
    a. place people in a particular status.
    b. educate children.
    c. serve as a passive servant of the stratification system.
    d. a and c
    e. none of the above

17. Research from "High School and Beyond" has found that:
   a. students in public schools are far more likely to drop out than are students in Catholic and private schools.
   b. white students are far more likely to do homework than are African-American students.
   c. students in public schools are far more likely to say they expect to go to college than are students in Catholic and private schools.
   d. all of the above
   e. none of the above

18. Collins has argued that:
   a. the expansion of higher education has created a "credential society."
   b. colleges impart training vital for the performance of many jobs that now demand a college degree.
   c. credentials serve as a way to control entry to many positions.
   d. a and c
   e. all of the above

19. Today women make up slightly more than _____ of the labor force in the United States and Canada.
   a. 25 percent
   b. 40 percent
   c. 55 percent
   d. 68 percent
   e. 75 percent

20. Long-term unemployment tends to be concentrated:
   a. in the Appalachian region of the United States.
   b. in the Pacific region of the United States and Canada.
   c. among minorities.
   d. a and c
   e. all of the above

Essay

1. A. Name two functions of schools. (knowledge)
   B. Discuss Meyer's theory of educational functions. (comprehension)
   C: Contrast Meyer's theory with allocation theories. (analysis)

2. A. Explain the concept of a knowledge economy. (comprehension)
    B. Show the interplay of education and occupation in a knowledge economy. (application)
    C. Contrast an industrial economy with a knowledge economy. (analysis)

3. Trace the history of the U.S. educational system.

4. Discuss the findings of either Coleman's study or Heyn's study.

5. Discuss the reasons why the quality of education seems to have declined in recent years.

6. In an attempt to improve the quality of education, some school districts are considering major changes in the school year. Some are opting for a shorter summer vacation with longer and more frequent breaks throughout the school year.

    Assume you have been appointed to a task force to study the feasibility of such a change. Given your knowledge of the research discussed in this chapter what advantages and disadvantages do you foresee resulting from such a change? How would such a change affect the families of the students? Likewise, what might be the impact of such a change on the local economy and the workforce? Do you feel you would favor or oppose this change? Why?

Answers

Completion

1.	more, higher	11.	disadvantaged
2.	greater	12.	poorer
3.	scientific management	13.	declined
4.	smarter	14.	little or no
5.	knowledge	15.	educational status
6.	unemployed	16.	Allocation
7.	achievement	17.	knowledge, personnel
8.	class interests	18.	less
9.	poor	19.	occupational prestige
10.	reading	20.	labor force

Multiple-Choice

1.	d	11.	b
2.	b	12.	c
3.	b	13.	a
4.	e	14.	d
5.	c	15.	e
6.	e	16.	d
7.	d	17.	a
8.	c	18.	d
9.	a	19.	b
10.	a	20.	d

CHAPTERS 13 TO 16

REVIEW AND SPECIAL PROJECT

Review

Chapters 13 through 16 have focused on the major social institutions: the family, religion, the political order, the economy, and education. They not only discussed the nature of these institutions but also emphasized changes in them.

Special Project

You might wish to investigate a specific change that you believe has occurred in one of these institutions during the past few decades. Possible topics for investigation might include the following:

1. The increase in the number of one-parent homes
2. The change in the divorce rate of a particular category of people (families with young children, people older than 50, and so on)
3. The increase (or decrease) in church attendance or membership in a specific denomination or sect
4. The increase (or decrease) in political participation (voting rates, registration rates) of a particular group (people younger than 30, African-Americans, and so on)
5. The increase (or decrease) in the number of women in a particular profession

To ascertain these changes, you will need to obtain data from 1940 or 1950 and then compare these statistics with more recent years. Sociologists often use the wealth of information available from Gallup Polls, census data, and government reports when studying trends and changes over time. Most of these statistics are readily available through your school or public library. (You may be surprised to see the wide variety of information that is available.) You will also want to investigate research journals for specific studies in your area of investigation. *Sociological Abstracts,* an index of articles published in major journals, is of great assistance in locating relevant articles.

Once you have obtained your data, you may want to further investigate these trends. You may wish to locate studies and theories that attempt to explain these changes within the context of general societal change.

CHAPTER 17

SOCIAL CHANGE AND MODERNIZATION

Overview

A discussion of modernization and sources of social change opens Chapter 17. It explains cultural lag, describes capitalism as an economic system, and contrasts capitalism with command economies. The chapter then offers an in-depth discussion of four theories of modernization: the Marxist view, Weber on Protestantism and the emergence of capitalism, the state theory of modernization, and the world system (dependency) theory. The chapter then turns its attention to tests of dependency theory by Delacroix and others. The chapter closes with a special topic on stirrups and feudal domination.

Capsule Summary

Modernization is the process by which agrarian societies are transformed into industrial societies. Social systems undergo change from both internal and external sources. Internal sources of change include innovations, new technology, new culture, new social structures, group conflict, and growth. External sources include diffusion, conflict, and ecological change. Cultural lag often occurs during times of social change.

Despite their differences, all theories of modernization attribute modernization in the West to capitalism. In contrast to command economies, capitalism is characterized by private ownership, competition for profits, and a free market. Capitalism encourages an individual to produce as much as possible because it rewards hard work and reinvestment of profits.

Karl Marx attributed the Industrial Revolution to capitalism. He believed that capitalism encouraged people to work harder and develop ideas, thus fostering techno-logical advances. Although he argued that capitalism was necessary for modernization, he believed that by promoting self-interest, capitalism fostered alienation, inequality, and class conflict. Marx believed that once modernization was accomplished, communist revolutions would foster collective ownership and allow the benefits of modernization to be equally shared.

Max Weber explained the emergence of capitalism as a result of religious doctrines of the Protestant Reformation. He argued that belief in predestination fostered an ideology that encouraged production, thrift, and the reinvestment of profit. Over time these values lost much of their religious significance and became basic secular values that were congruent with the economic system of capitalism.

The state theory of modernization argues that both capitalism and Protestantism are the result of the taming of the state. Repressive societies are characterized by command economies. As the power of the state is limited, people become freer to pursue economic self-interest; this in turn will encourage technological progress and capitalism. This theory argues that capitalism can only emerge once the state has been tamed.

In contrast to the preceding theories, <u>world system (dependency) theory</u> looks to <u>external</u> sources of change. It argues that, in the world system, stratification exists among nations. The <u>dominant</u> nations, termed <u>core nations</u>, exploit the weaker <u>peripheral</u> nations. <u>Core nations</u> are highly modernized while modernization in peripheral nations is hampered by this domination. <u>Research</u> by <u>Delacroix</u> and others has <u>not supported</u> the assumptions of <u>dependency theory</u>.

Key Concepts

You should be able to explain the concepts listed here and be able to give several examples of each.

Modernization 526	Capitalism 537
Innovation 526	Command economies 537
Cultural lag 530	Core nations 542
Social evolution 531	Peripheral nations 542
Diffusion 531	Semiperipheral nations 542

Key Research Studies

You should be familiar with both the methodology and the results of these studies.

<u>World Values Survey</u>: belief in scientific advances 528

Delacroix: test of the dependency theory 544

Firebaugh and Beck: test of dependency theory 547

Key Theories

You should be prepared to explain the assumptions of these theories of modernization and, when applicable, cite related research findings.

Karl Marx: capitalism

Max Weber: the Protestant Ethic and the emergence of capitalism

State theory of modernization

World system (dependency) theory

Completion

1. The process by which agrarian societies are transformed into industrial societies is termed _____.

2. Diffusion is the transfer of _____.

3. Types of innovation that may cause social change include new technology, new culture, and new _____.

4. The delay between the change in one part of society that produces a realignment of the other parts is known as _____.

5. The concept of cultural lag is associated with _____.

6. External sources of change include diffusion, conflict, and _____.

7. An economic system that is based on private ownership of the means of production and relies on a free market is termed _____.

8. A unique feature of capitalism is that it relies on a(n) _____.

9. Economies in which some people decide what work is to be done and order others to do it are termed _____.

10. The secret of capitalism is to reward _____.

11. _____ argued that the religious ideas produced by Protestantism motivated people to limit their consumption and pursue maximum wealth.

12. The state theory of modernization argues that _____ will always develop when the state is tame.

13. Chirot argued that the untamed state is incapable of not stifling economic development because it is incapable of not _____.

14. The doctrine of predestination is associated with _____.

15. _____ considers relationships among nations as a causative force in change.

16. Wallerstein argued that within the world system _____ exists among nations.

17. _____ nations have highly specialized economies, weak internal political structures, and a low standard of living for workers.

18. Wallerstein termed the dominant nations in the world system _____ nations.

19. Delacroix argued that modernization is influenced primarily by _____ processes.

20. Firebaugh and Beck have found that economic development in the Third World _____ the quality of life for everyone.

Multiple-Choice

1. Internal sources of social change include:
    a. innovations.
    b. group conflicts.
    c. growth.
    d. a and c
    e. all of the above

2. New technology:
    a. may appear and go unused for a long time.
    b. changes societies by itself.
    c. can be a major source of social change.
    d. a and c
    e. all of the above

3. Innovation may result in:
    a. new technology.
    b. new culture.
    c. new social structures.
    d. a and c
    e. all of the above

4. The periods of delay between the time one part of society changes and the other parts realign is termed:
    a. diffusion delay.
    b. cultural lag.
    c. innovation lag.
    d. cultural discontinuation.
    e. none of the above

5. The term *cultural lag* is associated with:
    a. Max Weber.
    b. Karl Marx.
    c. William Ogburn.
    d. John Calvin.
    e. Immanuel Wallerstein.

6. The transfer of innovations is termed:
   a. cultural lag.
   b. diffusion.
   c. accommodation.
   d. social evolution.
   e. assimilation.

7. External sources of social change include:
   a. diffusion.
   b. conflict.
   c. changes in the physical environment.
   d. a and b
   e. all of the above

8. An economic system based on private ownership of the means of production and a system by which people compete to gain profits is termed:
   a. capitalism.
   b. communism.
   c. socialism.
   d. privateering.
   e. a command economy.

9. Capitalism is *unique* in its:
   a. economic system based on private ownership of the means of production.
   b. emphasis on competition to gain profits.
   c. reliance on a free market.
   d. emphasis on communal goods.
   e. a and b

10. Command economies:
    a. rely on free-market principles.
    b. reward surplus production.
    c. encourage consumption.
    d. a and b
    e. all of the above

11. Capitalistic economies:
    a. rely on a free market.
    b. encourage immediate consumption.
    c. reward surplus production.
    d. a and c
    e. all of the above

12. The doctrine of predestination is associated with:
    a. Martin Luther.
    b. Karl Marx.
    c. John Calvin.
    d. Immanuel Wallerstein.
    e. none of the above

13. Weber argued that:
    a. capitalism blossomed from its roots in the Protestant Ethic.
    b. capitalism became a secular ideology in its own right.
    c. the Protestant Ethic was the sole cause of capitalism.
    d. a and b
    e. all of the above

14. The state theory of modernization argues that:
    a. capitalism will develop when the state is tame.
    b. Protestant theology led to the development of capitalism.
    c. modernization is the result of changes introduced from other societies.
    d. a and c
    e. all of the above

15. Theories that seek the causes of the Industrial Revolution *within* societies include the:
    a. world system theory.
    b. Marxist theory.
    c. dependency theory.
    d. b and c
    e. a and c

16. According to the world system theory, core nations have:
    a. weak or unstable governments.
    b. a low standard of living for workers.
    c. highly diversified economies.
    d. b and c
    e. all of the above

17.   In his test of the dependency hypothesis, Delacroix found that:
       a.   nations specializing in raw material exports showed as much an increase in
            per capita GNP as did nations specializing in the export of manufactured
            goods.
       b.   modernization is influenced by external processes of the world system.
       c.   extensive support exists for the dependency hypothesis.
       d.   all of the above
       e.   none of the above

18.   According to the world system theory:
       a.   stratification exists among nations.
       b.   the class position of a nation is determined by its place in a geographical
            division of labor.
       c.   less developed nations are economically dominated by more developed
            ones.
       d.   all of the above
       e.   none of the above

19.   Research by Firebaugh and Beck:
       a.   supported the dependency hypothesis by showing that economic
            development in the Third World benefits only the rich.
       b.   reaffirmed Delacroix's finding that investments in secondary education
            greatly speed economic development.
       c.   disproved Delacroix's finding that investments in secondary education
            greatly speed economic development.
       d.   a and b above
       e.   a and c above

*20.   In feudal societies, land ownership is based on _____ obligations.
       a.   economic
       b.   kinship
       c.   military
       d.   religious
       e.   none of the above

*Drawn from Special Topic 5.

Essay

1. A. Name the four theories of modernization. (knowledge)
    B. Explain two of these theories. (comprehension)
    C. Compare and contrast two of these theories. (analysis)

2. A. Define cultural lag. (knowledge)
    B. Give several examples of cultural lag. (comprehension)
    C. Apply this concept to the relationship of the introduction of pesticides to overpopulation in tropical nations. (application)

3. Discuss internal and external sources of social change and give an example of each.

4. Explain Weber's theory of the relationship between the Protestant Ethic and the emergence of capitalism.

5. Explain the world system theory and show how research has not supported the dependency hypothesis.

6. Social change is often the result of both internal and external factors. In recent years computerization has dramatically changed the nature of education, the economy, and other institutions. While the numerous benefits are obvious, such change can result in cultural lag and be problematic for certain segments of the population.

    If you are a traditional-aged college student, what changes have you noticed as a result of the increased use of computers in your lifetime? If you have younger siblings can you cite any differences between how your classes were conducted in elementary and high school and those they are experiencing? Do you find your parents or grandparents are less comfortable with this technology than you are? How might cultural lag explain this?

    If you are a returning adult, do you find differences between your attitudes and use of technology and those of the younger students? Again, how might you explain these differences? Despite the obvious benefits of these changes can you think of any disadvantages which might have resulted from this change?

Answers

Completion

1. modernization
2. innovations
3. social structures
4. cultural lag
5. William Ogburn
6. ecology
7. capitalism
8. free market
9. command economies
10. surplus production

11. Weber
12. capitalism
13. overtaxing
14. Calvin
15. World system theory
16. stratification
17. Peripheral
18. core
19. internal
20. improves

Multiple-Choice

1. e
2. d
3. e
4. b
5. c
6. b
7. e
8. a
9. c
10. c

11. d
12. c
13. d
14. a
15. b
16. c
17. a
18. d
19. b
20. c

CHAPTER 18

POPULATION CHANGES

Overview

Chapter 18 begins with a discussion of demography and early uses of government census. It then focuses on the various rates and measures used by contemporary demographers. After examining preindustrial population trends and Malthusian theory, it discusses population changes resulting from modernization and focuses on the theory of demographic transition. The chapter looks at the second population explosion and recent research on the fertility decline in developing nations. It closes with Special Topic 6, changes in society caused by the baby boom.

Capsule Summary

Demography is the study of population. Demographers study not only population size but also population changes and trends. Although demographic theory has its roots in the work of Adam Smith, ancient governments often conducted a census (such as the Domesday Book) for tax purposes. Demographers today often use extensive measures such as crude rates, specific rates, cohorts, and age and sex structures to measure population trends and provide a basis for long-term planning.

      Early societies often had difficulties in maintaining their populations. The first major increase in population occurred with the development of agriculture. Agricultural societies can produce more food and hence support greater numbers, but they are vulnerable to famines and disease, which serve to reduce their numbers. With the advent of modernization, a population explosion occurred as a result of both agricultural innovations and a marked decrease in the mortality rate.

      Malthusian theory attempts to explain the periodic growth and decline of populations prior to modernization. It postulates that populations always grow to a size slightly above the available food supply. Positive checks such as disease and famine then reduce the population to a size congruent with the available food, and the cycle begins again.

      The theory of demographic transition attempts to explain the population growth associated with modernization. This transition involves a change from the long-established pattern of high fertility and high but variable mortality to one of low fertility and low mortality. It argues that while modernization markedly cuts the death rate, it also encourages decreased fertility as large families become a liability rather than an asset. Cultural lag may occur between the initial decline in the mortality rate and the corresponding decline in the birth rate, causing a temporary rapid increase in population. This has occurred in the past few decades in less developed countries, although recent research has detected the beginning of a substantial decline in fertility (depopulation) in some of these nations.

Key Concepts

You should be able to explain the following concepts and be able to cite several examples of each.

Domesday Book 555
Demography 556
Growth rate 557
Crude death rate 557
Crude birth rate 557
Fertility rate 557
Age-specific death rate 558
Birth cohort 559
Age structure 560

Sex structure 560
Expansive population structure 560
Stationary population structure 562
Constrictive population structure 562
Positive checks 566
Numeracy 576
Wanted fertility 578
Depopulation 579

Key Research Studies

You should be familiar with both the methodology and the results of these research studies.

Data prepared by the author on fertility decline in less developed nations 573
Kristoff and Busch: low fertility and gender bias 581

Key Theories

You should be able to explain the assumptions of these theories and, when applicable, cite related research findings.

Malthusian theory
Demographic transition theory (Davis)
Berelson: "thresholds" of modernization

Completion

1. A population count is termed a(n) _____.

2. Demography is the study of _____.

3. The net population gain (or loss) divided by the size of population constitutes the _____.

4. The _____ can be computed by dividing the total number of deaths for a year by the total population for that year.

5. The total number of births divided by the total number of females within a certain age span is termed the _____.

6. All persons born within a given time period such as a year constitute the _____.

7. An expansive population structure is characteristic of current populations in _____ nations.

8. A declining population reflects a(n) _____ population structure.

9. The first great shift in population trends was caused by the development of _____.

10. Sudden rises in mortality rates in primitive societies can be the result of war, _____, and _____.

11. *An Essay on the Principles of Population* was written by _____.

12. Malthus termed famine, disease, and war to be _____.

13. The second great shift in population was caused by the _____.

14. _____ fertility occurs when the number of births each year equals the number of deaths.

15. The demographic transition theory is closely associated with _____.

16. As a result of modernization, children ceased to be a(n) _____ and became a(n) _____.

17. The fourth great shift in population trends was massive, unprecedented population growth in _____.

18. Recent data have shown a massive _____ in fertility in many less developed nations.

19. The term _____ is used to identify the capacity to use numbers.

20. The measure of _____ is the number of children a couple wishes to have.

Multiple-Choice

1.   The study of population is termed:
        a.   ecology.
        b.   ethnology.
        c.   democracy.
        d.   demography.
        e.   none of the above

2.   A population can decline because:
        a.   births are increasing.
        b.   deaths are increasing.
        c.   people are migrating into a region.
        d.   a and c
        e.   all of the above

3.   The number of deaths in a year divided by the total population for that year is termed the:
        a.   crude death rate.
        b.   general mortality rate.
        c.   age-specific death rate.
        d.   growth rate.
        e.   death cohort.

4.   All persons born in a given time period constitute the:
        a.   crude birth rate.
        b.   age-specific birth rate.
        c.   general fertility rate.
        d.   growth rate.
        e.   birth cohort.

5.   An expansive population structure:
        a.   reflects a declining population.
        b.   has fewer people at the bottom than in the middle.
        c.   is characteristic of underdeveloped nations.
        d.   a and b
        e.   all of the above

6. A rapid increase in the death rate may be the result of:
    a. famine.
    b. disease.
    c. war.
    d. all of the above
    e. none of the above

7. Which of the following has/have a positive effect on fertility?
    a. affluence
    b. religion
    c. the proportion of males in the population
    d. a and b
    e. all of the above

8. The first great shift in population trends was caused by:
    a. the modernization of agriculture.
    b. the development of agriculture.
    c. the decline in mortality due to better sanitation standards.
    d. the introduction of technology.
    e. none of the above

9. According to Malthus:
    a. population growth tends to rise slightly above the supply of food.
    b. fertility can be controlled through moral restraint.
    c. both fertility and mortality periodically rise and fall.
    d. a and c
    e. all of the above

10. According to Malthus, positive checks included:
    a. disease.
    b. moral restraint.
    c. war.
    d. a and c
    e. all of the above

11. During the initial period of modernization:
    a. the industrialization of agriculture increased the food supply.
    b. the population grew rapidly.
    c. the mortality rate dropped markedly.
    d. all of the above
    e. none of the above

12. Replacement-level fertility:
    a. occurs when the number of births each year equals the number of deaths.
    b. produces zero population growth as soon as the age structure has adjusted.
    c. has yet to be reached in industrialized nations.
    d. a and b
    e. all of the above

13. The demographic transition involves:
    a. a change from high fertility to low fertility.
    b. a change from low fertility to high fertility.
    c. a change from low mortality to high mortality.
    d. b and c
    e. none of the above

14. Davis argued that modernization encourages low fertility because:
    a. the decline in infant and childhood mortality eliminated the need for families to have large numbers of children to ensure that some survived to adulthood.
    b. large families became an asset rather than a burden.
    c. birth control devices were invented early in the Industrial Revolution.
    d. a and b
    e. all of the above

15. "Thresholds" of modernization include the characteristic(s) that:
    a. more than half the labor force is not employed in agriculture.
    b. 80 percent of the females age 15 to 19 are not married.
    c. at least one-half of the adults can read.
    d. a and b
    e. all of the above

16. The second population explosion:
    a. occurred in Western nations.
    b. occurred during the 1970s and early 1980s.
    c. resulted from a rapid drop in the mortality rate.
    d. a and b
    e. all of the above

17.   By the early 1970s, demographers detected a(n) _____ in the less developed
        nations.
        a. fertility increase
        b. fertility decline
        c. mortality increase
        d. mortality decline
        e. none of the above

18.   The proportion of males and females in a population constitutes the:
        a. sex rate.
        b. sex structure.
        c. sex ratio.
        d. fertility rate.
        e. fertility ratio.

19.   An age structure in which younger cohorts are smaller than the ones before
        them is termed a(n):
        a. stationary population structure.
        b. constrictive population.
        c. expansive population.
        d. modernized population.
        e. none of the above

20.   When populations are shrinking:
        a. the largest birth cohorts are always the oldest ones.
        b. the size of the labor force declines more rapidly than the size of the elderly
            population.
        c. shortages in the workforce can be offset through immigration.
        d. b and c above
        e. all of the above

Essay

1.   A. Define demography. (knowledge)
        B. Explain Malthusian theory. (comprehension)
        C. Contrast Malthusian theory with the demographic transition theory.
            (analysis)

2. A. Define the demographic transition theory. (knowledge)
    B. Using the concept of cultural lag, explain the second developing nations. (comprehension)
    C. Apply either the demographic transition theory or Malthusian theory to the current situation in developing nations. (application)

3. Describe the population characteristics of preindustrial societies.

4. Explain some of the causes and possible effects of "depopulation."

5. Discuss the changes that have occurred as a result of the baby boom. (Special Topic 6)

6. Demographers are often (but not always) able to predict future population trends. Assume you are considering a career in business, medicine, or education, in which the need for certain products and services will change as certain age cohorts increase or decline. How could you use these predictions to determine what areas of specialization in your chosen profession will be in demand in the future? Do you feel these areas could also be affected if parents are better able to choose the sex and birth order of their offspring? If so, what might be some of the consequences for your chosen area?

Answers

Completion

1.	census	12.	positive checks
2.	population	13.	Industrial Revolution (or modernization of agriculture)
3.	growth rate		
4.	crude death rate	14.	Replacement-level
5.	fertility rate	15.	Kingsley Davis
6.	birth cohort	16.	economic asset, economic burden
7.	underdeveloped	17.	less developed nations
8.	constrictive	18.	decline
9.	agriculture	19.	numeracy
10.	disease, famine	20.	wanted fertility
11.	Malthus		

Multiple-Choice

1.	d	11.	d
2.	b	12.	e
3.	a	13.	a
4.	e	14.	a
5.	c	15.	e
6.	d	16.	c
7.	b	17.	b
8.	b	18.	b
9.	a	19.	b
10.	d	20.	e

CHAPTER 19

URBANIZATION

Overview

After opening with a description of preindustrial cities, Chapter 19 describes the impact of the agricultural revolution and industrialization on urban growth. It introduces the concept of a metropolis and distinguishes between the fixed-rail metropolis and the freeway metropolis. Ethnic neighborhoods are then discussed, highlighting both Park and Burgess's early theory and recent tests of that theory. The chapter then examines the work of early theorists such as Tönnies, Durkheim, and Wirth and cites relevant research. It ends with a discussion of both the macro and micro effects of crowding.

Capsule Summary

Urbanization, the migration from rural areas to cities, was the result of modernization. Prior to the agricultural revolution and industrialization, cities were small, dirty, disease-ridden, and crowded. Despite these conditions, people migrated to preindustrial cities in search of economic gain, adventure, and anonymity. The agricultural revolution made it possible for larger numbers of people to live in the city; specialization also required a large urban work force.

Today the term city is rather nebulous. Because many people live in suburbs surrounding a city, the term metropolis (or metropolitan area) has come to refer to a city and its sphere of influence.

Modern cities have been shaped by transportation. Older industrial cities are termed fixed-rail metropolises because their growth followed the railroad lines outward from the center of the city. More recently, the decentralized freeway metropolis has emerged, and research indicates that people seem to prefer to reside in such an area.

Many cities contain ethnic and racial neighborhoods. The ethnic populations of these areas change over time by a process termed succession. Park and Burgess argued that slum neighborhoods are successively occupied by the lowest status groups. Guest and Weed, using the index of dissimilarity, recently found support for this theory and concluded that the barriers to integration were economic rather than ethnic or racial. Taeuber recently reported that U.S. cities have become less segregated as African-Americans have moved into the suburbs and white neighborhoods. Research by Farley and Frey confirmed the general trend towards integration although the inclusion of suburban areas caused some metropolitan areas to become more segregated. Studies of the residential patterns of Hispanic- and Asian-Americans have found less segregation than among African-Americans.

Early theorists took a negative view of the city. Tönnies's contrast of Gemeinschaft and Gesellschaft portrayed Gesellschaft relationships as cold and impersonal. Durkheim (and later Wirth) argued that urban areas were characterized by high rates of anomie and resulting deviance. However, recent research has failed to support the assumption that anomie is characteristic of urbanites.

Researchers have studied both macro and micro effects of crowding. Studies have not found any significant differences in pathology rates between areas of high and low population density. On the other hand, research by Gove and others found considerable support for micro effects of crowding.

Key Concepts

You should be able to explain the following concepts and be able to cite several examples of each.

Preindustrial cities (characteristics) 593

Specialization 600

Suburb 603

Metropolitan area (metropolis) 603

Sphere of influence 603

Standard Metropolitan Statistical Area 603

Fixed-rail metropolis 604

Freeway metropolis 606

Ethnic succession 608

Index of dissimilarity 609

Gemeinschaft 614

Gesellschaft 614

Anomie 614

Key Research Studies

You should be familiar with both the methodology and the results of these research studies.

Guest and Weed: economics and integration 609

Taeuber: test of Guest and Weed study 610

Farley and Frey: metropolitan integration and segregation 611

Denton, Massey, and others: Hispanic and Asian residential segregation 612

World Values Survey: desirable neighbors 613

Faris and Dunham: neighborhood affluence and rates of mental illness 615

Macro studies of crowding 617

Gove and others: micro study of crowding 618

Key Theories

You should be able to explain the assumptions of these theories and, when applicable, cite related research findings.

Park, Burgess, and McKenzie: theory of ethnic succession

Tönnies: *Gemeinschaft* and *Gesellschaft*

Durkheim and Wirth: anomie theories

Completion

1. The migration of people from the countryside to the city is termed _____.

2. Urbanization is the result of the more general process of _____.

3. The successive occupation of slum neighborhoods by the lowest status groups is termed _____.

4. Limits on the size of preindustrial cities included poor transportation and _____.

5. _____ and urbanization are inseparable processes.

6. An elaborate division of labor to simplify production is termed _____.

7. If an area has a population of more than 2,500, demographers classify it as a(n) _____.

8. The U.S. census classifies a community as a city when it has at least _____ residents.

9. A city and its suburbs with the central city as a single unit is termed a(n) _____.

10. The _____ of a city is the area whose inhabitants depend on the central city for jobs, recreation, and a sense of community.

11. The focal point of _____ cities was the center of the city.

12. Decentralized cities, common in the western United States, are termed _____.

13. The theory of ethnic succession was proposed by _____ and _____.

14. Park and Burgess proposed that ethnic and racial segregation in cities was based primarily on _____ and _____ differences.

15. Guest and Weed argued that _____ between groups seemed to be the primary neighborhood barrier.

16. Hispanic- and Asian-American residential patterns show _____ segregation than among African-Americans.

17. Gove and others found support for _____ rather than _____ theories of crowding.

18. Tönnies used the term _____ to describe small cohesive communities.

19. Tönnies used the term *Gesellschaft* to describe the quality of life in _____ societies.

20. An urban place in the immediate vicinity of a city is termed a(n) _____.

Multiple-Choice

1.   The size of preindustrial cities was limited by:
   a.   disease.
   b.   poor transportation.
   c.   reliance on nearby farms to provide food.
   d.   a and b
   e.   all of the above

2.   Today, approximately _____ of Americans and Canadians are urban residents.
   a.   25 percent
   b.   50 percent
   c.   60 percent
   d.   75 percent
   e.   90 percent

3.   Which of the following were reasons why people were drawn to preindustrial cities?
   a.   economic incentive
   b.   the prospect of a more interesting and stimulating life
   c.   the comparative safety of life in the city compared with that in the small town
   d.   a and b
   e.   all of the above

4.   Industrialization:
   a.   made it possible for most people to live in cities.
   b.   made it necessary for most people to live in cities.
   c.   depends on specialization.
   d.   all of the above
   e.   none of the above

5.   According to the U.S. census, to qualify as a city a community must have at least _____ residents.
   a.   2,500
   b.   20,000
   c.   50,000
   d.   75,000
   e.   100,000

6. If a locale has a population of more than 2,500, demographers term it a(n):
   a. urban place.
   b. urban area.
   c. community.
   d. city.
   e. village.

7. The area surrounding a city whose inhabitants depend on the central city for jobs, recreation, media, and a sense of community constitutes the city's:
   a. metropolitan area.
   b. sphere of influence.
   c. metropolis.
   d. zone of transition.
   e. zone of influence.

8. An area counts as a Standard Metropolitan Statistical Area if it:
   a. has a central city of 50,000 or more.
   b. is surrounded by a county in which 75 percent of those working in the county work in agriculture.
   c. is surrounded by a county in which 15 percent of the workers commute to the central city for work.
   d. a and c
   e. all of the above

9. Fixed-rail cities:
   a. made the center of the city the focal point.
   b. are common in the western United States.
   c. are more evenly spread out than freeway metropolises.
   d. all of the above
   e. none of the above

10. According to the Gallup Poll, most Americans preferred to live:
    a. in a city.
    b. on a farm.
    c. in a suburb or small town.
    d. b and c
    e. none of the above

11. The theory of ethnic succession is most closely associated with:
    a. Park and Burgess.
    b. Tönnies.
    c. Durkheim.
    d. Wirth.
    e. none of the above

12. The degree of segregation or integration of a neighborhood is measured by an:
    a. index of similarity.
    b. index of dissimilarity.
    c. index of status characteristics.
    d. index of ethnic characteristics.
    e. none of the above

13. Guest and Weed:
    a. argued that Park and Burgess were discussing individual upward mobility.
    b. argued that the status inequality between groups seems to be the primary neighborhood barrier.
    c. found evidence that would discredit the theory of Park and Burgess.
    d. a and b
    e. all of the above

14. Research by Farley and Frey found the highest levels of segregation in:
    a. retirement areas.
    b. areas dominated by large military bases.
    c. university communities.
    d. b and c above
    e. none of the above

15. Macro studies of crowding have found that:
    a. considerable support exists for the "psychic overload" theory.
    b. city people are more prone to alcoholism and mental illness than are rural people.
    c. neighborhoods with high population densities have much higher rates of pathology than do less dense neighborhoods.
    d. all of the above
    e. none of the above

16. The micro studies of Gove and others found that:
    a. there is little support for micro theories of crowding.
    b. people in crowded homes had poorer mental health.
    c. members of crowded homes had poorer mental health.
    d. b and c
    e. none of the above

17. Studies of Hispanic and Asian segregation have found that:
    a. the degree to which Hispanics and Asians are segregated declines very markedly with income.
    b. Hispanics and Asians are far more segregated than are African-Americans.
    c. segregation is higher for both Asians and Hispanics born in this country than for recent immigrants.
    d. b and c
    e. none of the above

18. Characteristics of *Gesellschaft* are apparent when:
    a. people are united only by self-interest.
    b. group members share little agreement about the norms, and deviance is common.
    c. human relationships are fleeting and manipulative.
    d. all of the above
    e. none of the above

19. Durkheim argued that:
    a. a primary consequence of urbanization was the breakdown of order.
    b. urbanites live in a state of anomie.
    c. rural areas have higher crime rates than urban areas.
    d. a and b
    e. all of the above

20. Research by Gove et al. on crowding found that:
    a. child care in crowded homes was poor.
    b. the effects of crowding began to show up only when there were more than two people per room in a household.
    c. people responded to crowding by withdrawing mentally and physically.
    d. a and c
    e. all of the above

Essay

1.   A.   Define urbanization. (knowledge)
        B.   Describe some characteristics of preindustrial cities. (comprehension)
        C.   Discuss the interplay between urbanization and modernization. (analysis)

2.   A.   Identify Park and Burgess's theory. (knowledge)
        B.   Explain this theory. (comprehension)
        C.   Show how recent research studies have (or have not) supported this theory. (analysis)

3.   Discuss the characteristics of preindustrial cities and show how these characteristics limited their size.

4.   Distinguish between Tönnies's concepts of *Gemeinschaft* and *Gesellschaft*.

5.   Discuss macro and micro theories of crowding and cite relevant research findings.

6.   Residential segregation seems to be on the decline in many areas although recent research has also indicated this has not been true in all areas.
How would you characterize the degree of integration in your hometown? Does it have specific areas which are more or less segregated than others? How would you characterize the specific area of the city or town in which you live?

If you grew up in this area, have you noticed any significant changes since you were a child? If you moved during your childhood was it to a more or less segregated area? Was the move in any way related to increasing integration? (Remember, as certain areas became more integrated some groups moved in while others moved out.)

If you attend college away from home does your college town reflect a greater or lesser degree of integration than your hometown? Does your own personal experience seem consistent with the research findings discussed in this chapter?

Answers

Completion

1. urbanization
2. modernization
3. succession
4. disease
5. Industrialization
6. specialization
7. urban place
8. 50,000
9. metropolis
10. sphere of influence
11. fixed-rail
12. freeway metropolises
13. Park, Burgess
14. economic, status
15. status inequalities
16. less
17. micro, macro
18. *Gemeinschaft*
19. industrial
20. suburb

Multiple-Choice

1. e
2. d
3. d
4. d
5. c
6. a
7. b
8. d
9. a
10. c
11. a
12. b
13. b
14. a
15. e
16. d
17. a
18. d
19. d
20. d

CHAPTER 20

THE ORGANIZATIONAL AGE

Overview

After describing the characteristics of formal organizations, Chapter 20 uses examples from the military, private business, and government to describe the process of centralization in nineteenth-century organizations. It then discusses Weber's concept of rational bureaucracy and the rational systems approach. This approach is contrasted with the natural systems approach to the study of formal organizations. The chapter next considers the process of decentralization in private business and highlights the theories of Blau and Thompson. It closes with a look at the increasing centralization in government.

Capsule Summary

Formal (or rational) organizations differ from older forms of organization in that they apply reason to the problems of management. Characteristics of formal organizations include a clear statement of goals, operating principles and procedures for pursuing these goals, trained leaders, clear lines of communication and authority, and written communication and records. During the nineteenth century, formal organizations emerged in such diverse areas as the military, private industry, and government.

　　Max Weber termed these organizations rational bureaucracies. His approach, often termed the rational system approach, emphasizes the official and intended characteristics of an organization. The natural systems approach, on the other hand, focuses on the processes of goal displacement, goal conflict, and informal relations among members. Rather than opposing one another, these three processes are actually complementary.

　　Although the nineteenth century saw the centralization of business and industry, more recently the trend in these organizations has been decentralization. This process relies on autonomous divisions, differentiation, and discretion. The theories of Blau and Thompson, which have focused on this process, have led to considerable empirical research.

　　Ironically, although private businesses have become increasingly decentralized, governments have tended to become even more centralized. When applied to government, the terms bureaucracy and bureaucrat have taken on negative connotations in the minds of many people. Because governments are not as vulnerable as private organizations, recent critics have suggested that they too should be subject to objective evaluations of performance.

Key Concepts

You should be able to explain the concepts listed here and be prepared to cite several examples of each.

Formal organization 624	Goal conflict 634
Rational organization 624	Span of control 638
Functional division 627	Autonomous division 639
Vertical integration 627	Decentralization 639
Spoils system 631	Management by objectives 640
Bureaucracy 631	Discretion 640
Goal displacement 633	

Key Theories

You should be able to explain the assumptions of these theories and, when applicable, cite related research findings.

Weber: rational system approach
Natural system approach
Blau: theory of organization
Thompson: decentralization, discretion, and coalition formation

Completion

1. The _____ organization applies reason to the problems of management.

2. During the _____ century, the first large formal organizations were created.

3. The political practice of giving public offices to the supporters of the winning politician is termed the _____.

4. For Weber, the term *bureaucracy* was inseparable from the term _____.

5. The rational system approach emphasizes the _____ and _____ characteristics of organizations.

6. The _____ systems approach emphasizes the informal and unintended characteristics of organizations.

7. _____ occurs when organizations change their goals in pursuit of survival.

8. _____ occurs when different groups within an organization tend to pursue different goals.

9. The _____ system approach argues that the overriding goal of organizations is to survive.

10. The limit on the number of people a given person can supervise effectively is termed the _____.

11. The key element in the decentralization of organizations is _____.

12. Blau argued that the _____ the organization, the greater the proportion of total resources that must be devoted to management.

13. Discretion involves both the _____ for making decisions and the _____ to carry them out.

14. The more serious the potential consequences of an error are perceived to be, the _____ willing people will be to assume discretion.

15. Recently, business has become more _____, while government has become increasingly _____.

16. The strength of private bureaucracies is their _____.

17. The organization of military troops into small, identical units, each containing all military elements, is an example of a(n) _____ system.

18. Parts of an organization, each of which includes a full set of functional divisions, are termed _____ divisions.

19. The dispersing of authority from a few central administrators to persons directly engaged in activities is termed _____.

20. A situation in which managers and subordinates agree on goals that subordinates will try to achieve is termed _____.

Multiple-Choice

1.   During the _____ century, formal organizations developed in the military, business, and government.
   a.   twentieth
   b.   nineteenth
   c.   eighteenth
   d.   seventeenth
   e.   sixteenth

2. Formal organizations differ from older forms of organization in that formal organizations:
    a. depend on a clear statement of goals.
    b. possess clear lines of authority and communication.
    c. use written records and communications.
    d. a and c
    e. all of the above

3. Rational bureaucracy was first described by:
    a. Peter Blau.
    b. Max Weber.
    c. James Thompson.
    d. Emile Durkheim.
    e. none of the above

4. In the spoils system:
    a. the benefits of public office go to the supporters of winning politicians.
    b. people are encouraged to make a career of government service.
    c. people are prevented from making a career of government service.
    d. a and b
    e. a and c

5. The _____ system approach emphasizes the informal and unintended characteristics of organizations.
    a. rational
    b. natural
    c. informal
    d. irrational
    e. unnatural

6. According to Weber, bureaucracy is based on:
    a. functional specialization.
    b. a blurring of lines of authority.
    c. managers promoted on the basis of the spoils system.
    d. a and b
    e. all of the above

7. Critics of the rational approach claim that:
    a. the real lines of communication in organizations are not always the same as those on the organizational chart.
    b. all members always pursue the same goals.
    c. this approach is limited.
    d. a and c
    e. all of the above

8. According to the natural system approach, the overriding goal of organizations is to:
    a. specialize.
    b. promote self-distrust.
    c. survive.
    d. diversify.
    e. a and d

9. The situation termed _____ may occur when different groups within an organization pursue different goals than those of the organization.
    a. goal conflict
    b. goal displacement
    c. decentralization
    d. functional integration
    e. none of the above

10. A company that is _____ controls each step in the process of bringing its products to the consumer:
    a. horizontally integrated
    b. decentralized
    c. vertically integrated
    d. functionally disintegrated
    e. none of the above

11. When the March of Dimes changed its focus from the elimination of polio to the elimination of birth defects, it exhibited:
    a. functional integration.
    b. goal displacement.
    c. goal conflict.
    d. horizontal integration.
    e. rationalization.

12. A major shortcoming of the rational systems approach is that it fails to emphasize:
   a. formal structure.
   b. goal displacement.
   c. the organizational blueprint.
   d. b and c
   e. all of the above

13. The limit on the number of people that a given person can supervise effectively is termed the:
   a. functional limit.
   b. span of control.
   c. vertical integration limit.
   d. supervisory sphere.
   e. none of the above

14. To survive, the DuPont Company had to institute:
   a. centralization.
   b. functional limits.
   c. autonomous divisions.
   d. supervisory spheres.
   e. all of the above

15. In his theory of administrative growth, Blau argued that:
   a. as organizations become less diversified, the size of administrative components increases relative to the size of other components.
   b. the smaller the organization, the greater the proportion of total resources that must be devoted to management function.
   c. organizational growth causes differentiation.
   d. a and c
   e. all of the above

16. Thompson argued that members of an organization will accept discretion when:
   a. they believe they cannot adequately control conditions affecting decisions.
   b. they do not share the responsibility of that decision with others.
   c. the decision involves forces outside the organization.
   d. a and c
   e. all of the above

17.   Which of the following statements is/are true?
    a.   As governmental organizations have grown larger, they have decentralized.
    b.   Centralization has dominated governmental organizations, while decentralization has dominated private organizations.
    c.   As governmental organizations have grown larger, they have become more centralized.
    d.   b and c
    e.   none of the above

18.   Breaking an organization into smaller units on the basis of specialized activities is termed:
    a.   decentralization.
    b.   the natural system approach.
    c.   geographical division.
    d.   functional divisions.
    e.   none of the above

19.   When the official goals of an organization are ignored or changed, it is termed:
    a.   goal conflict.
    b.   goal rejection.
    c.   goal replacement.
    d.   goal displacement.
    e.   none of the above

20.   The rational system approach emphasizes:
    a.   the intended characteristics of the organization.
    b.   the informal characteristics of the organization.
    c.   the unintended characteristics of the organization.
    d.   b and c
    e.   a and b

Essay

1.   A.   Name the two approaches to the study of bureaucracy. (knowledge)
    B.   Explain each of these approaches. (comprehension)
    C.   Contrast these approaches. (analysis)

2.   A.   Define formal organization. (knowledge)
    B.   Describe four official and intended characteristics of formal organizations. (comprehension)
    C.   Apply these characteristics to the military, government, and private industry. (application)

3. Describe the characteristics of formal organizations that distinguish them from older forms of organization.

4. Using the examples in the text, trace the development of formal organizations in the military, private business, and government.

5. Explain the process of decentralization. Discuss the theory of *either* Blau *or* Thompson on the decentralization process.

6. Today colleges and universities are bureaucratic organizations and, as such, have both rational and natural systems.

    Assume that upon entering college you learn that the official procedure for course selection and registration is conducted on the basis of seniority; that is, seniors are allowed to register first while freshmen must register last. This obviously gives upperclassmen the advantage in enrolling in certain courses and sections while freshmen frequently find desired classes closed. After a semester or two, however, you notice some students, especially those who "volunteer" to assist with registration or who are friendly with certain faculty or staff manage to circumvent these rules and register early. How would you use your knowledge of both rational and natural systems to explain this?

    If you feel this is unfair, what steps could you take to help remedy the situation? If you wanted to improve your chances for a better schedule, how might you use the natural system to achieve your ends?

Answers

Completion
1. formal or rational
2. nineteenth
3. spoils system
4. rationality
5. official, intended
6. natural
7. Goal displacement
8. Goal conflict
9. natural
10. span of control
11. discretion
12. larger
13. responsibility, authority
14. less
15. decentralized, centralized
16. vulnerability
17. divisional
18. autonomous
19. decentralization
20. management by objectives

Multiple-Choice

1.	b	11.	b
2.	e	12.	b
3.	b	13.	b
4.	e	14.	c
5.	b	15.	c
6.	a	16.	b
7.	d	17.	d
8.	c	18.	d
9.	a	19.	d
10.	c	20.	a

CHAPTER 21

SOCIAL CHANGE AND SOCIAL MOVEMENTS

Overview

Using the Civil Rights Movement as an example, Chapter 21 focuses on how individuals can create change through social movements. Many of the themes of this text are illustrated in the discussion of the birth and evolution of this movement. The chapter contrasts the collective behavior and the resource mobilization approaches and offers a synthesis of the two. It then discusses the birth of the Civil Rights Movement and focuses on the Montgomery Bus Boycott and the actions taken by leaders such as Martin Luther King, Jr. Internal and external factors of social movements are examined within this context. The Freedom Summer Project is described, and McAdam's study of participants of this project closes the chapter.

Capsule Summary

Social movements occur whenever people organize to cause or prevent social change. Social movements are often studies from either the collective behavior approach or the resource mobilization approach. The collective behavior approach emphasizes the importance of grievances, emotions and feelings, and the role of ideology. The resource mobilization approach, on the other hand, emphasizes the importance of human and material resources and the role of rational planning. Neither approach seems to fully explain social movements, and a synthesis is necessary to more fully explain the birth and evolution of a social movement.

For a social movement to occur, some members of the society must share a grievance that they want to correct. They also must have hope that change is possible. (A revolution of rising expectations occurs when people come to believe that change may occur rapidly. The J-curve theory of social crisis illustrates the relationship between expectations and objective conditions.) A precipitating event is needed to trigger action. Once begun, however, a social movement needs to effectively mobilize people and resources, withstand or overcome external opposition, and enlist external allies if it is to succeed.

The Civil Rights Movement in the American South is a classic example of a successful social movement. Before its inception, African-Americans shared grievances about economic, political, and personal domination by whites. Changes were, however, taking place that led to hope and rising expectations. When Rosa Parks refused to give up her seat on a bus, the Montgomery Bus Boycott began and the Civil Rights Movement was born.

Once begun, African-American churches provided the internal factors of effective leadership, communication among participants, and physical facilities such as meeting places free from white domination. Through the efforts of leaders such as Dr. Martin Luther King, Jr., external allies were recruited and the movement gained strength and support. As is typically the case, whites opposed to the movement began a countermovement that led to harassment and, ultimately, violence. The Freedom Summer

Project recruited volunteers from <u>white</u>, <u>affluent backgrounds</u> who were already politically active in civil rights causes. (Research has shown that twenty years later most of these volunteers maintained attachments within the liberal community.) Violence against these volunteers led to media attention and drew national support and recognition of the movement.

Key Concepts

You should be able to explain the concepts listed here and be able to give several examples of each.

Social movement 649	Internal factor 656
Countermovement 650	External factor 657
Revolution of rising expectations 653	Biographical availability 664
Precipitating event 654	

Key Research Study

You should be familiar with both the methodology and the results of this study:

McAdam: Freedom Summer study 664

Key Theories

Be prepared to explain the assumptions of the following theories and, when applicable, cite research findings.

Collective behavior approach
Resource mobilization approach
Synthesis of collective behavior and resource mobilization (eight propositions)
J-curve theory of social crisis

You should also be familiar with each person and be able to identify his or her role in the Civil Rights Movement.

Rosa Parks
Martin Luther King, Jr.
JoAnn Robinson

Completion

1. Whenever people organize to cause or prevent social change, we identify them as a(n) _____.

2. Sociologists committed to the collective behavior approach emphasize social movements as outbursts of group activity in response to deeply felt _____.

3. The collective behavior approach also places great significance on the role of _____ in fixing the goals and tactics of social movement.

4. The resource mobilization approach stresses the importance of _____ and _____ as both a source of movements and a basis of their success.

5. The resource mobilization perspective places its greatest emphasis on _____.

6. For a social movement to occur, a(n) _____ must ignite pent-up grievances and convince people that the time for action has arrived.

7. Usually a social movement will generate one or more _____.

8. Internal factors that influence a social movement include securing the necessary finances and facilities, effective leadership, and _____.

9. A(n) _____ occurs when people come to believe that things can be made better rapidly.

10. The _____ the gap between expectations and reality, the more impatient and frustrated people become.

11. A dramatic event that galvanizes people into action is termed a(n) _____.

12. The primary _____ factor of a social movement is a group of people with sufficient commitment and motivation to engage in activities designed to cause change.

13. The primary _____ factor is the degree to which the movement can be suppressed by those opposed to its actions.

14. African-Americans in the south prior to the Civil Rights Movement suffered from political, _____, and _____ domination by whites.

15. The J-curve theory of social crisis predicts that there is apt to be a severe crisis when there is an actual _____ in what people are receiving.

16. Rather than acting in a vacuum, Rosa Parks stood at the center of a(n) _____ that could serve as a basis for recruiting people to form a social movement.

17. The early evolution of the Civil Rights Movement was based on the only institutions over which African-Americans had total control, their _____.

18. The _____ project recruited white volunteers from elite colleges and universities to register African-American voters in Mississippi.

19. McAdam found that people volunteered for Freedom Summer on the basis of _____ to members of civil rights organizations.

20. McAdam discovered that the major impact of participation on the volunteers in Freedom Summer was to make them more politically _____.

Multiple-Choice

1. The collective behavior approach to social movements:
   a. emphasizes social movements as outbursts of group activity in response to deeply felt grievances.
   b. places great significance on the role of ideology in fixing goals and tactics.
   c. stresses the importance of rational decision making on the part of participants.
   d. a and b
   e. all of the above

2. The resource mobilization perspective:
   a. places its greatest emphasis on leadership.
   b. places its greatest emphasis on grievances as the basis of social movements.
   c. ignores the role of rational planning.
   d. all of the above
   e. none of the above

3. According to the author, for a social movement to occur:
   a. some members of a society must share a grievance that they want to correct.
   b. the members must have hope that there is some possibility of success.
   c. a precipitating event must ignite grievances and convince people that the time for action has arrived.
   d. a and c
   e. all of the above

4. For a social movement to succeed:
   a. it must achieve an effective mobilization of people and resources.
   b. it must withstand or overcome external opposition.
   c. it must enlist external allies from other organizations or at least keep them neutral.
   d. a and c
   e. all of the above

5. Which of the following is the correct sequence of events necessary for a social movement to occur?
   a. hope, grievance, precipitating event
   b. grievance, precipitating event, hope
   c. precipitating event, grievance, hope
   d. grievance, hope, precipitating event
   e. precipitating event, hope, grievance

6. The author argues that sociologists committed to the resource mobilization approach:
   a. pay too little attention to the realities of grievances expressed by movements.
   b. pay too little attention to the organizational realities involved in social movements.
   c. focus exclusively on the grievances voiced by social movements.
   d. b and c
   e. none of the above

7. A social movement will be more successful when it:
   a. enjoys effective leadership.
   b. is able to secure necessary finances and facilities.
   c. attracts allies from other groups and institutions within societies.
   d. a and b
   e. all of the above

8. Internal factors of social movements include:
   a. effective leadership.
   b. suppression by those opposed to their aims.
   c. material resources.
   d. a and c
   e. all of the above

9. The primary internal factor of a social movement is:
   a. a group of people with sufficient commitment and motivation to engage in activities designed to cause change.
   b. effective leadership.
   c. the degree to which the movement can be suppressed by those opposed to its aims.
   d. a means of communication among members.
   e. material resources.

10. The shared grievances experienced by African-Americans in the South before the Civil Rights Movement included:
    a. economic domination.
    b. political domination.
    c. personal domination.
    d. all of the above
    e. none of the above

11. A revolution of rising expectations occurs when:
    a. people feel there is no hope for change.
    b. people believe that things can be made better slowly.
    c. people believe that things can be made better rapidly.
    d. there is no difference between what people get and what they hope to get.
    e. a and d

12. According to the J-curve theory of social crisis, a crisis is most apt to occur when:
    a. what people hope to get and what they actually get are relatively close.
    b. expectation for change and rate of change both rise rapidly.
    c. the rate of change suddenly declines.
    d. the rate of change suddenly increases.
    e. a and b

13. "A dramatic event that galvanizes people into action" is termed a(n):
    a. precipitating crisis.
    b. precipitating event.
    c. event horizon.
    d. revolution of rising expectations.
    e. action mobilization.

14. Economic domination of African-Americans by whites in the South before the Civil Rights Movement included:
    a. help-wanted ads listing "for whites only."
    b. whites deciding who would be hired, fired, or promoted.
    c. higher rates of employment among white women than among African-American women.
    d. a and b
    e. all of the above

15. When Rosa Parks refused to give up her seat on a Montgomery bus:
    a. she acted on impulse.
    b. she had a network of attachments within the African-American community.
    c. she believed that many members of the African-American community shared her perception that the time for resistance had come.
    d. b and c
    e. all of the above

16. African-American churches were especially effective internal factors of the Civil Rights Movement because:
    a. African-American ministers were skilled and experienced organizers.
    b. they could provide meeting places and had the ability to raise money.
    c. they were the only institutions over which Southern African-Americans had total control.
    d. a and b
    e. all of the above

17. McAdam discovered that people who became volunteers in Freedom Summer:
    a. often had no previous contact with civil rights organizations.
    b. were linked to the project by preexisting social relationships.
    c. typically were married and held full-time jobs.
    d. all of the above
    e. none of the above

18. Freedom Summer volunteers were characterized by:
    a. a strong sense of personal power.
    b. a strong belief in racial equality.
    c. a rebellious attitude against their parents' conservative beliefs.
    d. a and b
    e. all of the above

19.   McAdam's research found that participants and no-shows could be
       distinguished on the basis of:
       a.   network ties.
       b.   attitudes.
       c.   values.
       d.   b and c
       e.   all of the above

20.   Research on Freedom Summer participants twenty years later revealed that:
       a.   most have lost their idealism and tend to be politically conservative.
       b.   they are less likely than others of their age group to be married.
       c.   most have been employed at the same job for a long period of time.
       d.   all of the above
       e.   none of the above

Essay

1.   A.   Name the two approaches used to study social movements.
             (knowledge)
      B.   Explain the assumptions of each approach. (comprehension)
      C.   Apply each approach to the Civil Rights Movement. (application)

2.   A.   Name the three conditions necessary for a social movement to occur.
             (knowledge)
      B.   Explain these conditions in-depth. (comprehension)
      C.   Show how these conditions were met in the birth of the Civil Rights
             Movement. (application)

3.   Trace the development of the Civil Rights Movement in the 1950s and 1960s.

4.   Discuss some of the results of McAdam's research on Freedom Summer
      participants. Show how attachments have played a role in their political
      participation.

5.   Explain the following statement: "To the question 'Do we make our history or
      does it make us?' the answer clearly is 'Yes!'"

6.   While many people associate the Civil Rights Movement with the 1950's,
      60's, and early 70's, this movement remains active today. The current
      controversy surrounding affirmative action for minorities and women has its
      roots in this movement. Likewise other current social issues such as abortion
      rights and gay rights also trace their heritage to this movement.

What current movements are active in your college community or hometown? Given your knowledge of the Civil Rights Movement, Freedom Summer, and McAdam's research, what categories of people would you predict would be active on either side of these issues?

If possible, you might attend a meeting or rally in support or against one of these issues and interview those present. Were your results consistent with your expectations or did your findings surprise you?

Answers

Completion

1.	social movement	11.	precipitating event
2.	grievances	12.	internal
3.	ideology	13.	external
4.	resources, rational planning	14.	economic, personal
5.	leadership	15.	decline
6.	precipitating event	16.	network of attachments
7.	countermovements	17.	churches
8.	attracting committed and disciplined members	18.	Freedom Summer
9.	revolution of rising expectations	19.	attachments (personal ties)
10.	greater	20.	radical

Multiple-Choice

1.	d	11.	c
2.	a	12.	c
3.	e	13.	b
4.	e	14.	d
5.	d	15.	d
6.	a	16.	e
7.	e	17.	b
8.	d	18.	d
9.	a	19.	a
10.	d	20.	b

CHAPTERS 17 TO 21

REVIEW AND SPECIAL PROJECT

Review

Chapters 17 through 21 focused on changes brought about by modernization. They examined population trends, urban growth and development, the centralization and decentralization of organizations, and social movements.

Special Project

Social movements such as the Civil Rights Movement often trigger major social change. The impact of this change is felt not only by those directly affected by the movement (or countermovement) but also by all segments of society.

Since its birth in the late 1950s, the Civil Rights Movement has had a major impact on our society, although changes in population and technology and numerous other factors have intertwined to foster this change.

You may wish to examine the impact of some of these changes in your college or university, your hometown or neighborhood, your place of employment, or even your family. Be alert to both subtle and more obvious changes. For example, you may wish to examine one or more of the following:

1. Demographic changes in the composition of the students, faculty, or administration of your school.
2. Changes in curriculum requirements at your school.
3. Demographic changes in the composition of employees at your workplace.
4. Changes in requirements and job descriptions of positions in your workplace.
5. Changes in residential patterns within your neighborhood or city.
6. Changes in voting behavior and political participation within your area.
7. Changes within your family such as educational and occupational aspirations, an increase in two-income families, divorces and remarriages, family size, and so on.

Possible sources of information include demographic data from administrative offices, census tracts, personal interviews, as well as more unobtrusive measures as newspapers, yearbooks, employee manuals, old photographs, and/or family albums. You might wish to formulate and test specific hypotheses or you might choose to conduct a more open-ended exploratory study. In either case, you may be in for some interesting surprises.